Applied Satellite Navigation Using GPS, GALILEO, and Augmentation Systems

For a listing of recent titles in the *Artech House Mobile Communications Series*, turn to the back of this book.

Applied Satellite Navigation Using GPS, GALILEO, and Augmentation Systems

Ramjee Prasad

Marina Ruggieri

ARTECH
HOUSE

BOSTON | LONDON
artechhouse.com

Library of Congress Cataloging-in-Publication Data
Prasad, Ramjee.
    Applied satellite navigation using GPS, GALILEO, and augmentation systems / Ramjee Prasad, Marina Ruggieri.
        p. cm. — (Artech House mobile communications series)
    Includes bibliographical references and index.
    ISBN 1-58053-814-2 (alk. paper)
    1. Artifical satellites in navigation   I. Ruggieri, Marina.   II. Title.   III. Series.
    VK562.P73   2005
    623.89'3–dc22

                                                                    2005041995

British Library Cataloguing in Publication Data
Prasad, Ramjee
    Applied satellite navigation using GPS, GALILEO, and augmentation systems. — (Artech House mobile communications series)
    1. Artifical satellites in navigation   2. Global positioning system
    I. Title   II. Ruggieri, M. (Marina), 1961–
    629'.045
    ISBN 1-58053-814-2

Cover design by Yekaterina Ratner

International Standard Book Number: 1-58053-814-2

10 9 8 7 6 5 4 3 2 1

To our children Anand, Flavia, Luca, Neeli and Rajeev
and grandchildren Akash, Ruchika, and Sneha

Contents

CHAPTER 10
Open Issues and Perspectives   253

Preface

मुक्तसंगोऽनहंवादी धृत्युत्साहसमन्वितः ।
सिद्ध्यसिद्ध्योर्निर्विकारः कर्ता सात्त्विक उच्यते ॥२६॥

mukta-saṅgo 'nahaṁ-vādī
dhṛty-utsāha-samanvitaḥ
siddhy-asiddhyor nirvikāraḥ
kartā sāttvika ucyate

One who performs his duty without association with the modes of material nature, without false ego, with great determination and enthusiasm, and without wavering in success or failure is said to be a worker in the mode of goodness.

—*The Bhagvad Gita* (18.26)

Before May 1, 2000—when U.S. President Bill Clinton ordered the removal of the intentional performance degradation of the Global Positioning System (GPS)—satellite-based navigation was considered a niche of the entire space market and military forces were its main users.

That day dramatically changed the destiny of satellite navigation: in fact, the increased accuracy for civilian GPS receivers, induced by Clinton's decision, has opened the door to an avalanche of novel services, new typologies of users, and advanced integration architectures for both existing systems and newly conceived ones.

The satellite navigation world is hence experiencing a quick and fascinating evolution; and if we believe that a book is a sign of a time worthy of being described in a permanent form, this is certainly the time to write a book on satellite navigation. The question is, which book about navigation do we currently need?

GPS and its modernization plan seem mature for several civilian applications; the U.S. Department of Defense is studying the new generation system GPS III; Europe, through a joint venture between the European Commission and the European Space Agency, is living the fascinating challenge of developing GALILEO, whose preliminary phases have been completed. In the meantime, thanks to the success of GPS, navigation users' awareness about the incredible potential of Global Satellite Navigation Systems (GNSS) continuously increases. In this context, the deployment of satellite-based augmentation systems in various countries (e.g., WAAS in the United States and Canada, EGNOS in Europe, MSAS in Japan, GAGAN in India, SNAS in China), as well as ground- and aircraft-based augmentations, is somehow filling the gap between the navigation users' dreams and the time-to-deployment of the second generation GNSS (i.e., GALILEO and GPS III).

Another interesting aspect in the evolving picture of the navigation world concerns integration with communications systems: wireless network designers have become fully aware of the importance of localization for advanced services and they are ready to implement satellite navigation technologies in new generation user terminals.

These activities highlight the need for a book focused on future trends of satellite navigation, taking a broad picture of the past, capturing the present evolution and challenges, and envisaging the directions for an effective exploitation of navigation services in the user-centric global vision of the future.

This book provides a deep technical, scientific, and strategic vision of navigation systems belonging to different countries and in various development stages. The authors aimed at focusing with the same depth on the technical/architectural aspects and user-related aspects (terminals, services, security, guarantees) of satellite navigation, harmonizing them with legal/market issues as well as with hints and thoughts for new proposals and initiatives in the field.

Due to its multipurpose nature, this book can be used at various educational levels, including masters of science and doctoral students who might find it beneficial as a reference/design book and prospective source. At the same time, this book is suited for professional engineers and managers, as an aid in consolidating a future vision about the navigation world and in the subsequent planning and strategic decision-making process.

The reader is offered a large view of the navigation world in the book's 10 chapters, which move from the navigation historical and analytical fundamentals (Chapters 1 and 2, respectively) through the description of present and future milestones of the satellite navigation scenario—GPS (Chapter 3), augmentation systems (Chapter 4), GALILEO (Chapter 5), GPS modernization and GPS III (Chapter 6)—up to key aspects of future trends in navigation, including legal and market policy (Chapter 7), services (Chapter 8), and integration with existing and future systems (Chapter 9). The book concludes with open issues and perspectives regarding the fascinating navigation adventure (Chapter 10).

Acknowledgments

This book has been realized thanks to the efforts and enthusiasm of an extraordinary team that deeply believes in the satellite navigation world and its potential. In particular, the authors wish to thank Daniele Teotino, Marino Donarelli, and Mirko Antonini for their valuable contribution.

The support of Cinzia Foglia and Junko Prasad is also gratefully acknowledged.

CHAPTER 1

Introduction

Navigation is the process that drives a person during his or her movement between two points, enabling an ancient dream of the human race: the knowledge of position at any given time. This need derives from the spirit of discovery and exploration that every human being has, a sign of our common essence:

> Considerate la vostra semenza:
> fatti non foste a viver come bruti,
> ma per seguir virtute e canoscenza.

> Think of the roots from which you sprang, and show
> that you are human, not unconscious brutes
> but made to follow virtue, and to know.

> —Dante Alighieri, *La Divina Commedia, Inferno, Canto XXVI*

We have maps that show the Earth, we have conquered the space around the Earth, and we have conquered the Moon; we are working to conquer the rest of the universe.

We cannot live without journeying and each journey, be it either only a few kilometers long or to where none has been before, needs support to reach the right location.

1.1 Introduction to Navigation: A Historical Perspective

Terrestrial navigation has relied for millenniums on maps, signs, and the human sense of direction [1, 2]; it is only in recent times that radio signals have helped land navigation. The most important tools for navigation were created to ensure safety during maritime journeys.

The first infrastructure created as a navigational aid was the lighthouse. The world's first lighthouses were erected about 2,000 years ago: the Colossus on the Rhodes Island in Greece and Pharos at Alexandria in Egypt. It was not until the nineteenth century that the Fresnel lens improved the performance of lighthouses by increasing their range significantly.

Sea travel was considered unsafe for a long time and the best-known explorers, such as Bartholomeu Diaz, Ferdinand Magellan, Sir Francis Drake, Amerigo Vespucci, and Christopher Columbus, were considered brave sea captains [3]. These men used primitive instruments to calculate the position and speed of their ships [4]. The mind of any sailor was closed to the new (for that age) astronomical studies for determining the latitude. Christopher Columbus crossed the Atlantic Ocean without any accurate measurement of the latitude. The main instruments used for calculating the latitude were the quadrant, astrolabe and nocturnal. Columbus tried to use these tools but their results were disappointing. All these devices aimed to measure the altitude of a celestial body above the horizon. The quadrant was the first instrument developed for celestial navigation, known back to the fifteenth century. It had the form of a quarter circle with degree graduations along the arc. The astrolabe was similar to the quadrant. This device was suspended from a cord to hang perpendicular to sea level, while the navigator sighted the sun or a star through two small holes in the plates on its moveable vane. The nocturnal, or night disc, consisted of two concentric circles of different sizes, made of either wood or brass, and it was also used to calculate the time at night. All these instruments were influenced by the ship's movement, which, in case of bad weather conditions, rendered measurements difficult. Columbus tried to use the quadrant latitude reading for the first voyage only on October 30, 1592, in Cuba, but the measurement showed him to be 42° north latitude instead of about 20°. He blamed the quadrant for the bad result and remarked that he would not take any more readings until the quadrant was fixed. Recent studies have demonstrated that Columbus read the wrong scale of the instrument, which was actually performing quite satisfactorily!

This could be considered a prime example of useful service implemented into a user-unfriendly terminal. Recent history has confirmed that users need services while, at the same time, they indeed do not need complex or interfering technologies.

Columbus and most of the sailors of his era navigated by deduced (or dead) reckoning method. This method was based on the measurement of the course and the distance from some known points. The course was measured by a magnetic compass, which was used in Europe since at least the twelfth century. The distance was measured every hour through the measurement of the speed. This was measured by throwing a piece of flotsam over the side of the ship; there were two points of the ship rail marked (one near the prow and one near the stern). The sailor started a quick chant (part of an ancient oral tradition of medieval navigation) when the flotsam passed the forward mark and stopped the chant when it passed the aft mark. Each syllable reached was equivalent to a speed in miles an hour. At the end of the day, the total distance and the course, calculated as the sum of the distance and course measured each hour, was written on the chart.

The dead reckoning method could be considered an ancestor of the modern inertial navigation systems (INS). INS works with instantaneous and continued measurement of speed, distance, time, and direction and it allows a user's knowledge of a new position each time when starting from a known one. It was initially developed for submarine navigation but is used nowadays for surface water, air, and land navigation as well. Because dead reckoning navigation does not need any astronomical observation, inertial navigation does not need any radio navigational aid.

A quadrant, if well used, could measure the latitude with an accuracy of about one degree, but at that time no instruments were used to measure longitude in a consistent and sufficiently accurate way. The problem of the longitude was keenly felt in the United Kingdom and the most important scientists of the day started to meet in Grisham College (London) to "find out the longitude":

> The Colledge will the whole world measure,
> Which most impossible conclude,
> And Navigators make a pleasure
> By finding out the longitude.
> Every Tarpalling shall then with ease
> Sayle any ships to th'Antipodes.
> —Anonymous, about 1661

In 1662, the scientists previously mentioned found the Royal Society of London for the Promotion of Natural Knowledge, even if the problem was not solved immediately. Between 1690 and 1707, several ships were lost during sea navigation due to the wrong estimation of their position. In 1707, after a 12-day storm, Admiral Sir Cloudesley Shovel evaluated wrongly to be west of the French islands of Ouessant and, as a result, four war ships struck rocks near the Scilly Isles and more than 2,000 men were killed. In 1714, Queen Anne established the "Board of Longitude," composed of scientists and admirals, to examine the proposals and check the results by accurate tests, by the Act of Parliament 12th, which declared as follows:

> That the first author or authors, discover or discoveres of any such method ... shall be entitled to, and have such reward as herein after is mentioned; that is to say, to a reward or sum of ten thousand pounds, if it determines the said longitude to one degree of a great circle, or sixty geographical miles; to fifteen thousand pounds if it determines the same to two thirds of that distance; and to twenty thousand pounds, if it determines the same to one half of that same distance... [5]

It is quite clear that such a vast amount of money (current value at more than $10 million) provoked the interest of scientists and alchemists in the longitude problem [6]. It is beyond the scope of this book to list the scientific or imaginative methods proposed to win the prize [7]. However, while the board expected an astronomical solution to the problem, it was solved through different means: longitude can be computed measuring the difference in time between midnight (or local moon, the highest point of the sun) and the midnight (or the moon) of a known place (e.g., Greenwich). The earth performs one full rotation (360°) every 24 hours, thus it rotates 15° every hour. This means that if the local moon is one hour before the Greenwich moon, the place is sited 15°W. Local moon or midnight time could be determined, respectively, by sundial and nocturnal [8, 9].

This method was well known and proposed by Gemma Frisius in 1530; unfortunately time on board ships was measured in a very inaccurate way by a sandglass, and a member of the crew had the responsibility to turn the glass (about every half hour). Clock artists grew fully in that age, but as Isaac Newton stated:

> One is, by a watch to keep time exactly: but by reason of the motion of the ship, the variation in heat and cold, wet and dry, and the difference in gravity at different latitudes, such a watch had not yet been made.

To win the full prize, the clock used (more specifically a chronometer) would have had to measure the absolute time to an accuracy of 2 minutes, all journey long.

> I should then see the Discovery of the Longitude, the perpetual Motion, the Universal Medicine, and many other great Inventions brought to the utmost Perfection.
> —*Gulliver's Travels*, Jonathan Swift, 1726

William Whinston and Humphry Ditton proposed a sort of optical transoceanic telegraph [10]: the method was to use a number of lightships anchored in the principal shipping lane at regular intervals. They would have transmitted the signal of the Greenwich midnight time by firing a star shell; these synchronization signals would then have propagated ship by ship. Although this method was impractical for several technical reasons, nowadays we can recognize a number of technologies that use this concept, starting from radio bridges and the frequency shortwave broadcast radio stations for radio clocks synchronization to very high frequency omni range (VOR) and distance measuring equipment (DME) stations for air navigation.

Where scientists, astronomers, and mathematicians failed, a carpenter's son with no formal education and self-taught craftsmanship won the longitude prize [5].

Between 1730 and 1759, John Harrison built four clocks that were the most accurate and precise timekeepers ever seen at that time. The last one, called H4, was a pocket-size watch, just 13 cm in diameter and weighing 1.45 kg. In a voyage to Barbados aboard the Tartar on March 28, 1764, the watch error was only 39.2 seconds over a journey of 47 days (about three times better than the standards required to win the full prize).

It is amazing how history has repeated itself many times. There is no need to cite the relativistic theory to understand the strict connection between space and time. The design of a global navigation system was conditioned by the availability of very accurate clocks, much more accurate than any common quartz clock. Nowadays everyone can know where he or she is to a few meters accuracy with the help of a GPS receiver. We take this for granted because atomic clocks that are precise to a billionth of a second are installed onboard the GPS satellites [11]. The concept of navigation is based on the knowledge of reference points. Just as celestial navigation, based on triangulation method, used celestial bodies as known points, many centuries later GPS, based on the concept of trilateration, uses satellites as known points, whereas long range radio aid to navigation (LORAN) uses fixed stations as known points.

1.2 Evolution of Satellite Navigation

An illuminating approach to delineate the evolution of satellite navigation systems is drawing the impact of this evolution in key areas of human life: history, politics, social, and economics.

This approach is hereinafter followed to analyze the milestones of satellite navigation development: GPS, GLObal'naya NAvigatsionnaya Sputnikovaya Sistema (GLONASS), and GALILEO.

1.2.1   The Pioneers

1.2.1.1   Global Positioning System

There is little doubt that satellite navigation is, in the mind of most people, synonymous with GPS. Satellite navigation was born when the U.S. Navy decided in the early 1960s to create a system for precise navigational purposes. The system was called Navy Navigation Satellite System or, more commonly, TRANSIT. The Navy submarine fleet could not use inertial navigation for a long time without periodic and accurate updates of their position; nevertheless, the main application that convinced the U.S. government of the need for funding such a system was to take accurate fixes for programming the launches of submarine launched ballistic missiles (SLBM). The system was developed by the Johns Hopkins Applied Physics Laboratory (APL) and ceased operation in 1996, made obsolete by GPS. Although this system, based on a Doppler shift measurement of a tone broadcast at 400 MHz, was very different from GPS, TRANSIT has proved the importance and necessity of navigation systems for the U.S. Army. In 1964, the Navy research laboratory Naval Centre for Space Technologies (NCST) created the TIMATION (TIMe/navigATION) program. Two TIMATION satellites were developed and launched in 1967 and 1969, carrying accurate quartz oscillators that were regularly updated by a master clock on the ground. These satellites opened the way to GPS because the user position was determined by passive ranging. In fact, the TIMATION and the U.S. Air Force project called 621B, which had developed the satellite ranging signal based on pseudorandom noise, gave birth to the NAVSTAR, for NAVigation Satellite Timing And Ranging, GPS. In the meantime the U.S. Department of Defense (DoD) decided to include all these programs in a single strategy, called Defense Navigation Satellite System (DNSS) under the control of a Joint Program Office (JPO) located at the Air Force Space and Missile Organization in El Segundo, Calif. Under the TIMATION program, two more satellites called NTS I, for Navigation Technology Satellite, and NTS II were launched in 1974 and 1977, respectively [12, 13]. They were the world's first satellites carrying atomic clocks, a rubidium and cesium one in turn, and they could be considered prototypes of GPS satellites. Therefore the NAVSTAR GPS program went ahead and in February 1978 the first GPS satellite was launched. During this time, all these programs were under the absolute control of the military forces and the motto that drove JPO action was as follows [14]:

> The mission of this Program is to:
> 1. Drop 5 bombs in the same hole, and
> 2. Build a cheap set that navigate (< $10,000),
> And don't you forget it!

Starting from 1978, GPS satellite launches were carried out continuously until the period between 1986 and 1989 [15] when DoD decided to revise its programs after the Space Shuttle Challenger accident in 1986 [16, 17]. The first generation of GPS satellites was composed of 12 satellites; 11 of them were successfully launched, one was destroyed due to launch failure, and one was never launched. This constellation, indicated as Block I, was launched from 1978 to 1985.

The second generation (Block II), composed of nine satellites, was launched with an intensive program from 1989 to 1990. The first major test for GPS was without doubt the Persian Gulf crisis in 1990–1991 that is recorded as "The Gulf War." About 16 active GPS satellites drove Operation Desert Storm; the military used a number of commercial units (more than 10,000) hastily ordered during the crisis that allowed a positioning accuracy of about 30m. The technical and tactical supremacy enabled by GPS is clear reading the reports of that time:

> "NAVSTAR GPS played a key role and has many applications in all functional war-fighting areas. Land navigation was the biggest beneficiary, giving Coalition forces a major advantage over the Iraqis."
> —Department of Defense Report to Congress, *Conduct of the Persian Gulf War*, Vol. II, April 1992

Launches of new satellites did not stop during the Gulf War, but the second evolution stage of the second generation of NAVSTAR GPS satellites was put in orbit between 1990 and 1997 [18]. Some of the 19 launched satellites of this block (Block IIA) are still active at the time of this book. In 1993, initial operational capability (IOC) was formally declared but the constellation of 24 satellites in six orbital planes (Block I and Block II/IIA satellites) was completed after a few months in 1994. In 1995, full operational capability (FOC) was declared with 24 Block II/IIA satellites launched. Table 1.1 summarizes the launch history of GPS satellites. The current generation of satellites launched (Block IIR, Replenishment) is composed of 21 elements overall, 12 of them already successfully launched between 1997 and 2004 (tycho.usno.navy.mil/gpscurr.html), one lost due to launch failure in 1997 and some converted to Block IIR-M (Modernized). The latter introduces a new military code (M-code; see Chapters 3 and 6). The M-code will enable military operations in high jamming environment. Starting in 2006, the launch of the fourth evolution stage of Block II satellites (Block IIF, Follow-on) is foreseen. The block will be composed of 12 satellites.

Current generations and future generations of GPS satellites (GPS III) will be discussed in Chapters 3 and 6, respectively. Although GPS was conceived for military purposes, since 1993 (the year IOC was declared), the U.S. Department of Transportation (DOT) and DoD have worked together on a joint task force to ensure the maximum civil use of GPS. In fact, it can be considered that the civil use of GPS technologies started in 1993 [19]. The signal available to civil user terminals was always corrupted by an intentional degradation, called *selective availability* (SA). The level of accuracy with SA was limited to 100-m horizontal position (95% probability) and 140-m vertical position (95% probability) [20]. The common experience of navigation services suggests that such accuracy was inadequate for most applications conceived until now [21]. The choice to implement the SA was dictated by the assumption that the enemy could use GPS as a weapon against the United

Table 1.1 Summary of the GPS Satellites Launches

GPS Satellites Block	Years	Successful Launches	Failed Launches
Block I	1978–1985	10	1
Block II	1989–1990	9	0
Block IIA	1990–1997	19	0
Block IIR	1997–2004	12	1

States [22, 23]. Modernization of GPS satellites and antijamming techniques demonstrated that SA was becoming an obsolete method. In fact, in case of crisis, DoD can order the "selective deniability" to nonauthorized users on a regional basis, according to Jeremy Eggers, a spokesman at Schriever Air Force Base in Colorado. "We have demonstrated the ability to selectively deny GPS signals on a regional basis...when our national security is threatened," he said. Therefore, on May 1, 2000, President Clinton stated:

> The decision to discontinue selective availability is the latest measure in an ongoing effort to make GPS more responsive to civil and commercial users worldwide...This increase in accuracy will allow new GPS applications to emerge and continue to enhance the lives of people around the world.

1.2.1.2   GLObal'naya NAvigatsionnaya Sputnikovaya Sistema (GLONASS)

GLONASS is historically framed, as in the case of GPS, in the development of technological tools for the Cold War between the United States and the former Soviet Union in the 1960s. The Russian GLONASS was very similar to GPS and, in spite of satellite launches started in the 1980s (the first GLONASS launch occurred on October 12, 1982), the system was officially declared in operation on September 24, 1993, IOC, under the decree of the president of the Russian Federation [24–28].

In the second half of 1996, the International Civil Aviation Organization (ICAO) and International Maritime Organization (IMO) officially accepted the proposal of the Russian Federation government to provide the world civil community with a standard-accuracy service through the GLONASS. In fact, besides providing users with navigation service, global navigation systems are excellent instruments for precise timing. Timing users worldwide tribute great importance to GPS and GLONASS use for precise universal time coordinated (UTC) transfer. Interest in capabilities and advantages of combined GLONASS/GPS use for this purpose was expressed in Recommendation S4 of the 13th session of Consultative Committee for Definition of Second (CCDS), dated March 13, 1996 [29, 30].

In spite of the large potential, the GLONASS system was never brought to completion, probably due to a combined lack of necessary funding, strong motivations after the end of the Cold War, solid perspectives for civilian applications, and related market opportunities.

The mission Coordinational Scientific Information Center (KNITs) undertakes planning, management, and coordination of the activities related to military (in particular, GLONASS) and civil space systems. GLONASS operates under the control of the Russian Federation Ministry of Defense.

At present, 11 GLONASS satellites are operational, hence, much less than those envisaged for the full constellation deployment; four of the operational satellites were launched in the first term of 2003. A twelfth satellite was launched in December 2004. A high uncertainty currently characterizes the system's future.

1.2.2   GALILEO: The European Challenge

Particularly in the last 15 years, noticeable progress in the European aerospace industry has been reported in several areas, such as manufacturing, launchers, management, control, and services. This growth has also rendered the European "product" competitive in trans-European projects, as in the development of the International Space Station.

A key step for the challenging evolution scenario of European technology and capability is the development of the GALILEO navigation system, a joint initiative of the European Space Agency (ESA) and the European Union (EU) (i.e., the institutions that represent the core of the European strategic, technological, and space-related policy).

GALILEO, moving from ideas and concepts conceived in the 1990s, will be a civilian system under international civil control. Within EU, GALILEO is seen as a transportation infrastructure, controlled by the General Directorate for Energy and Transport. In addition, the system has been opened to extra-European participation, such as China and India.

Europe has never been involved autonomously in such a large project as GALILEO. The design of such a fascinating and complex system needs a harmonized integration of many expertise and technologies, including those not fully related to the space world.

GALILEO is, hence, a key challenge from a technological viewpoint and it could help the European industry become competitive in all segments of the value chain. Investigations and projections indicate that more than 100,000 people will be employed thanks to GALILEO-induced benefits [31].

GALILEO will bring further social benefits. In the future, the knowledge of position could become as necessary as time knowledge is nowadays. In this respect, the transportation system will improve its efficiency and security thanks to an advanced navigation system such as GALILEO. Innovative and improved services will be rendered available to citizens [32–34]. For instance, navigation aids and up-to-date information about optimal routes will be available as transportation infrastructure, bringing a reduction in road traffic congestion and a consequent positive impact on air pollution.

At the end of 2000, the GALILEO program completed the system feasibility and definition phase, financed by the European Commission (EC) in accordance with the Trans European transport Network (TEN) program with 100 million euros. During that phase, the EC and ESA involved most of the European space industry, as well as potential service operators, to define major design elements.

The GALILEO development phase started in 2001 and is still ongoing [35–41]. Activities envisage the consolidation of mission requirements, the development of ground and space segments, and the implementation of the first satellite prototypes. This phase is being funded by the EC for 450 million euros from the TEN program and by the ESA for 550 million euros, through a joint venture of public and private funding that represents an important step toward the development of GALILEO activities and services. In fact, on May 2001, the GALILEO Joint Undertaking (JU) was created by the EC and ESA [42]. The JU manages, coordinates, and funds all GALILEO development activities for 4 years, with the aim of attracting private funding into the financial structure of the joint venture and, hence, easing system development (Figure 1.1). The challenging opportunities for manufacturers, space operators, and, in general, service operators in the context of the European space industry have allowed the

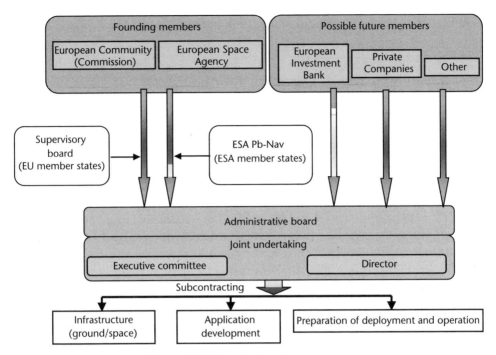

Figure 1.1   Organization of the GALILEO Joint Undertaking.

early aggregation of partners in consortia that favored the JU creation and implementation.

The JU memorandum of association established that the European Investment Bank and any private enterprise can participate in the JU through a minimum share of 5 million euros; medium and small enterprises can participate for 250,000 euros.

In 2006–2008, the deployment (construction and launch) of the GALILEO 30-satellite constellation is envisaged, together with implementing the ground infrastructure. The related costs will be supported by the JU, envisaging a share of about 600 million euros from public funding and 1.5 billion euros from private ones. The project's cost until the operation phase is estimated in 3.2 billion euros, whereas the recurring cost from the beginning of the operation phase is estimated at about 220 million euros per year. Added-value services are expected to compensate for the management costs. The GALILEO operational phase, IOC, is envisaged starting from 2008.

On October 20, 2003, the JU published a call for tenders to deploy the selection procedure of the GALILEO Concessionaire, which takes place in two stages: short-listing and   competitive negotiation. The timely completion of the concession process is a key step for the successful deployment of the GALILEO system and, hence, the full development of the envisaged market, whose figures are impressive.

The market already amounts to 10 billion euros per year; it is growing at a rate of 25% per year and will reach some 300 billion euros in 2020. More than 3 billion receivers are likely to be in service by 2020 [31]. The anticipated revenues are arousing strong interest among investors, hence the reason for choosing to administer the system in the form of a concession as part of a public-private partnership.

The GALILEO system will impact any type of transportation.

Road transportation will benefit mainly from the superstructure that GALILEO will create for users' management and information. The integration of a communication channel into the navigation message will allow info-mobility services independently on terrestrial networks. A GALILEO receiver will, hence, be capable of the information provision to users, allowing their "intelligent" navigation toward optimal routes. Global coverage and format standardization—at least at the European level—will allow users to move abroad without risking any loss in information contents.

Real-time information about traffic conditions or parking reservations could be centralized in either public or private institutions, increasing efficiency of the road transportation system.

Information about public and private transportation media (trains, flights, ships, buses) could be distributed to users through GALILEO local components. Local authorities could then integrate an information network for traffic management that, through officially certified information, will allow the development of info-mobility as public utility service. To this respect, the GALILEO user terminal could become—starting in 2008—a necessary equipment for road vehicle safety. Info-mobility services could also provide information about cinema and theatre performances, sports results, meteorological forecasts, finances, and so on. Furthermore, the exploitation of GALILEO could allow fleet-management services. Additional benefits related to the exploitation of GALILEO features and potentials that enhance citizen safety include investigation of who is responsible for car accidents or infringements, emergency management in natural disasters, civil protection operations, and route management of dangerous goods as well as military or police operations.

Rail transportation could benefit from the European coverage and features of GALILEO, for localization and procedure automation, that would reduce risks related to human-error-based disasters. Regarding road transportation, the centralized management of a train fleet plays a key role in enhancing the service for quality and user safety. GALILEO will also impact effectively on the efficiency of rail-based goods transportation.

Benefits will extend as well to maritime and fluvial transportation, easing route identification and change and improving ship-docking operations. The automated identification system will interface with the GALILEO positioning capability, enhancing safety of maritime transportation. GALILEO could also provide a significant impulse toward developing intermodal transportation, recognizing again the key role of the fluvial component in good transportation.

In addition, GALILEO could be exploited for high-accuracy localization of European satellites, as GPS is already being used in several newly conceived platforms.

GALILEO will also play a strategic role in various areas of air transportation. In fact, air traffic control issues are moving EU members toward common strategies for rationalizing the control points distribution and task integration over the European sky (Single European Sky) [43–45]. In this context, the exploitation of modern satellite localization technologies contributes significantly to the ongoing standardization process. Air space is an important but limited resource that needs to be managed wisely with full support from emerging technologies. The consequent improvement will impact dramatically in meeting flight time scheduling, optimizing air and ground personnel and, hence, in improving flight fares and travelers' safety.

The better the accuracy of the navigation system, the wider its role could become in assisting various flight phases. To this respect, GALILEO could become an essential technology in the air transportation system.

The picture of GALILEO that has been discussed is key to understanding the high expectations that the GALILEO experience is providing in the public and private sections of European technical, political, and social life (Figure 1.2). The technical means that will allow GALILEO to achieve the highlighted role and goals will be discussed in Chapter 5.

1.3   Preview of the Book

The book is organized in 10 chapters.

In this chapter, the evolution of navigation systems has been depicted through historical, political, social, and economical developments, providing the reader with interpretation keys for the technical choices that will be discussed in detail in the next chapters.

In Chapter 2, the basics of navigation are discussed, highlighting in particular the required navigation performance and navigation equations.

GPS is described in Chapter 3, which details the system's various segments, its signal characteristics, performance, and security issues, as well as enhancements envisaged in the modernized GPS approach.

In Chapter 4, the connection ring between pioneer systems such as GPS, and future networks like GALILEO is discussed, that is, the augmentation system concept (the so-called 1.5G of navigation systems). The main systems of this type are highlighted as well as their major features that improve first generation (GPS, GLONASS) performance.

Chapter 5 is dedicated to the European system GALILEO and its several innovative characteristics, services, unexplored potentials, and security issues.

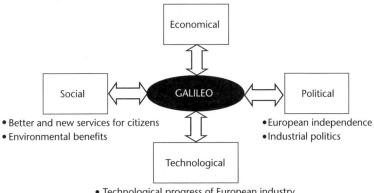

Figure 1.2   The four arguments of GALILEO: economical, political, social, and technological.

Moving from the modernized GPS concept, Chapter 6 brings forward the future generation of GPS, denominated GPS III, and its features.

Chapter 7 discusses the legal and market policies of satellite navigation in the context of commercial, public utility, and safety-of-life applications.

Chapter 8 presents an original layer-based classification of navigation services. The layers concern the medium where the navigation services are provided (i.e., Earth, sea, air, and space). Commonality and peculiarity among the layers are highlighted.

Chapter 9 addresses a key topic for the successful deployment of future navigation systems: the integration with existing and future communications systems. This includes the integration of wireless third generation (3G) and beyond 3G wireless systems.

Chapter 10 highlights open issues, possible solutions to current questions, and future perspectives for developing navigation systems in a harmonized and integration-prone vision of the communication world.

References

[1] Wilford, J. N., *The Mapmakers*, New York: Vintage, 2001.

[2] Brown, L. A., *The Story of Maps*, Boston, MA: Little, Brown, 1949.

[3] Armstrong, R., *The Discoverers*, New York: Time Life Science Library, 1966.

[4] Howse, D., and N. Maskelyne, *The Seaman's Astronomer*, New York: Cambridge University Press, 1989.

[5] Sobel, D., *Longitude: The True Story of a Lone Genius Who Solved the Greatest Scientific Problem of His Time*, New York: Penguin, 1996.

[6] Dash, J., and D. Petricic, *The Longitude Prize: The Race Between the Moon and the Watch-Machine*, New York: Frances Foster Books, 2000.

[7] Chapin, S., "A Survey of the Efforts to Determine Longitude at Sea," *Navigation*, Vol. 3, 1953, pp. 1660–1760.

[8] Forbes, E. G., "The Scientific and Technical Bases for Longitude Determination at Sea," *NTM Schr. Geschichte Natur. Tech. Medizin*, Vol. 16, No. 1, 1979, pp. 113–118.

[9] Howse, D., *Greenwich Time and the Discovery of the Longitude*, Oxford, England: Oxford University Press, Oxford, 1980.

[10] *Il cielo dei navigatori (The Sky of Navigators)*, the Astronomy Disclosure Committee and the Italian Astronautical Society, Florence, CD-ROM, Vol. 27, No. 9518, 1998.

[11] Taubes, G., *The Global Positioning System: The Role of Atomic Clocks*, Washington, D.C.: National Academy of Sciences, 2003.

[12] Lasiter E. M., and B. W. Parkinson, "The Operational Status of NAVSTAR GPS," *Journal of Navigation*, Vol. 30, 1977.

[13] Easton, R. L., "The Navigation Technology Program," *Institute of Navigation*, Vol. I, Washington D.C., 1980, pp. 15–20.

[14] Parkinson B. W., and J. J. Spilker Jr., *Global Positioning System: Theory & Applications*, Vol. I, American Institute of Aeronautics and Astronautics; 1st edition, January 15, 1996.

[15] Baker P. J., "GPS in the Year 2000 and Beyond," *Journal of Navigation*, Vol. 40, 1987.

[16] Parkinson B. W., "Overview, Global Positioning System," *Institute of Navigation*, Vol. I, Washington D.C., 1980, p. 1.

[17] Burgess, A., "GPS Program Status," *Proc. Nav 89: Satellite Navigation*, Royal Institute of Navigation, 1989.

[18]   Rip, M. R., and J. M. Hasik, *The Precision Revolution: GPS and the Future of Aerial Warfare*, United States Naval Institute, April 10, 2002.

[19]   Ho, S., "GPS and the U.S. Federal Radionavigation Plan," *GPS World*, Vol. 2, 1991.

[20]   *U.S. National Aviation Standard for the Global Positioning System Standard Positioning Service*, U.S. Department of Transportation, August 16, 1993.

[21]   Beser, J., and Parkinson B. W., "The Application of NAVSTAR Differential GPS in the Civilian Community," *Global Positioning System, Institute of Navigation*, Vol. 2, 1984.

[22]   Baker, P. J., "GPS Policy," *Proc. of the Fourth International Symposium on Satellite Positioning*, University of Texas at Austin, 1986.

[23]   Pace, S., et al., *The Global Positioning System: Assessing National Policies*, Santa Monica, CA: Rand Corporation, 1995.

[24]   Lebedev, M., et al., *GLONASS as Instrument for Precise UTC Transfer*, Warsaw, Poland: EFTF, March 1998.

[25]   Smirnov, Y., et al., *GLONASS Frequency and Time Signals Monitoring Network*, Warsaw, Poland: EFTF, March 1998.

[26]   Contreras, H., "GPS+GLONASS Technology at Chuquicamata Mine, Chile," *ION-98*, Nashville, TN, September 1998.

[27]   Coordinational Scientific Information Center of the Russian Space Forces, *GLONASS Interface Control Document*, 4th revision, 1998.

[28]   Coordinational Scientific Information Center of the Russian Space Forces, *GLONASS Interface Control Document*, 5th revision, 2002.

[29]   Allan, D. W., "Harmonizing GPS and GLONASS," *GPS World*, Vol. 5, 1996, pp. 51–54.

[30]   Coordinational Scientific Information Center of the Russian Space Forces, *GLONASS Interface Control Document*, October 4, 1995.

[31]   *Inception Study to Support the Development of a Business Plan for the GALILEO Programme, Final Report*, PriceWaterhouseCooper, November 14, 2001.

[32]   Antonini, M., et al., "Broadcast Communications Within the Galileo Commercial Services," *Proc. GNSS 2003—The European Navigation Conference*, Graz, Austria, April 2003.

[33]   Antonini, M., R. Prasad, and M. Ruggieri, "Communications Within the Galileo Locally Assisted Services," *Proc. IEEE Aerospace Conference*, Big Sky, MT, March 2004.

[34]   Ruggieri, M., and G. Galati, "The Space System's Technical Panel," *IEEE System Magazine*, Vol. 17, No. 9, September 2002, pp. 3-11.

[35]   *2420th Council Meeting—Transport and Telecommunications*, Brussels, March 25–26, 2002.

[36]   "Commission Communication to the European Parliament and the Council on GALILEO," Commission of the European Communities, Brussels, November 22, 2000.

[37]   "Council Resolution of 5 April 2001 on GALILEO," *Official Journal C 157*, May 30, 2001, pp. 1–3.

[38]   *Progress Report on GALILEO Programme*, Commission of the European Communities, Brussells, December 5, 2001.

[39]   "Communication from the Commission to the European Parliament and the Council: State of Progress of the GALILEO Programme," *Official Journal C 248*, October 10, 2002, pp. 2–22.

[40]   "Communication from the Commission to the European Parliament and the Council: Integration of the EGNOS Programme in the GALILEO Programme," COM (2003) 123 Final, March 19, 2003.

[41]   "Proposal for a Council Regulation on the Establishment of Structures for the Management of the European Satellite Radionavigation Programme," COM (2003) 471 Final, July 31, 2003.

[42]   "Council Regulation (EC) No 876/2002 of 21 May 2002, Setting Up the GALILEO Joint Undertaking," *Official Journal L 138*, May 28, 2002, pp. 0001–0008.

[43] "Communication from the Commission to the Council and the European Parliament,"
      *Action Programme on the Creation of the Single European Sky, Commission of the Euro-
      pean Communities*, Brussels, November 30, 2001.

[44] *Single European Sky, Report of the High Level Group*, European Commission, General
      Directorate for Energy and Transportation, November 2000.

[45] Iodice, L., G. Ferrara, and T. Di Lallo, "An Outline About the Mediterranean Free Flight
      Programme," *3rd USA/Europe Air Traffic Management R&D Seminar*, Naples, Italy, June
      2000.

CHAPTER 2

Navigation Basics

Current generation navigation systems (GPS and GLONASS) and those under development (e.g., GALILEO) determine the user terminal position through the time of arrival (TOA) ranging.

In general, this kind of ranging technique is based on the measurement of the time interval employed by a signal transmitted by an emitter (e.g., satellite, radio beacon) at a known location to arrive at the user receiver.

The TOA is defined by (2.1):

$$\text{TOA} = \text{Time Instant of Arrival} - \text{Time Instant of Transmission} \tag{2.1}$$

which is measured by the user receiver.

If the receiver knows the speed of the signal, it is able to determine the distance from the emitter simply by multiplying the TOA with the signal speed value. In the case of satellite navigation, electromagnetic signals, propagating at the speed of light (approximately 3×10^8 m/s), are used; therefore, the fundamental equation of satellite navigation is [1]:

$$\text{Speed of Light} \times \text{Time of Arrival} = \text{Distance} \tag{2.2}$$

From (2.2) it is clear that the calculation of the true (i.e., geometric) distance between the satellite and the receiver can be obtained only through the measurement of the true TOA, which implies, as highlighted by (2.1), that the receiver has a precise knowledge of the time instant of arrival and the time instant of transmission of the satellite signal. The former can be achieved through direct reading of the receiver clock, whereas the latter is embedded in the signal (see Chapter 3 for more details), which is nominated *navigation signal*. To achieve the true difference between these time instants, the satellite and receiver clocks have to be synchronized to the same time scale.

Once the user receiver has a sufficient number of distance values from multiple satellites with known locations, it can achieve its position, according to theoretical considerations drawn afterwards. In satellite-based navigation, transmitters are not fixed points, as in the terrestrial case. The receiver has, hence, to be able to determine the satellite position for each distance measurement. For this purpose, each navigation signal modulates a message that includes the satellite orbital parameters (known as *satellite ephemeris*) and, thus, enables the receiver to propagate the satellite orbit and, then, to evaluate the transmitter position at each time instant.

Orbital parameters are updated by the master control center (see Chapter 3) and transmitted to the satellites once or twice a day, but it is foreseen to increase the rate of uploading the satellites, also achieving a level of accuracy better than 10 cm [2] (see Chapter 6 for information about GPS modernization). At the time of writing, institutions such as the International GPS Service for Geodynamics (IGS) are able to provide precise ephemeris data to users with accuracies reaching even 1 dm [3].

The determination of a position needs the choice of a reference coordinate system, where both the satellites and the user receiver can be represented.

Furthermore, the acceptance of a unique time system for both satellite and receiver clocks arises to satisfy the necessity of time synchronization between them. For the following considerations it is sufficient to know that a time reference, the *GPS system time*, exists (see Section 2.2 for more details). This time reference is the basis for the operation of the whole satellite system composed of space components (satellites) and ground components (control centers and user receivers). GPS and GLONASS satellites use atomic clocks, which are accurate within one billionth of a second; however, size, mass/weight, and cost of atomic clocks render their use prohibitive at the user receiver. Therefore, an offset between the time measured at the satellite and at the receiver clocks is unavoidable, with a consequent error in the ranging measurement.

Another source of error is due to the differences between the GPS system time and the satellite clock time: information used to correct these discrepancies is uploaded by the GPS control segment (see Chapter 3) to satellites and then broadcast from the satellites as part of the navigation signal to user receivers to allow them to compute the opportune corrections [4].

Besides the errors due to time offsets between GPS system time and satellite clock time and between satellite clock time and receiver clock time, the ranging measurement is corrupted by other factors, namely, incorrect values of the satellite ephemeris, ionospheric and tropospheric navigation signal delay, receiver noise, and multipath [4, 5]. For a detailed treatment of these error sources, see Chapter 3.

All these reasons make the range measurement not coincide with the true (i.e., geometric) distance between the satellite and user receiver, hence causing one to designate such a range measurement *pseudorange*. Chapter 3 describes the technique adopted for the pseudorange measurement from the GPS navigation signal.

The ranging technique adopted in satellite navigation systems is *one way* and it has the advantage of serving simultaneously an unlimited number of users without requiring any transmission from their receivers (i.e., receivers are passive and not active).

This chapter focuses on the fundamentals of satellite navigation that should enable the reader to appreciate the implementation issues encountered in existing and future navigation systems. In particular, the chapter deals with the main coordinate systems and time systems, the concept of trilateration, the navigation equations and related solution techniques, together with the relatively new concept of specifying the requirements for navigation referred to as the required navigation performance (RNP).

Further subjects, including satellite orbit determination, ionospheric and tropospheric error modeling, multipath effects modeling, study of relativistic effects on

GPS, and radio-frequency interference on satellite signals and their mitigation techniques, are outside the scope of this book, so the authors recommend readers seek related information from [4–6].

2.1   Coordinate Systems

As already mentioned, the satellite navigation equations can be formulated only after having established a reference coordinate system valid for both the satellites and receiver. Generally, satellites and receiver are individuated by position and velocity vectors in a Cartesian coordinate system [5].

Before giving an overview of the coordinate systems adopted in satellite navigation, a few explanations are mandatory.

For sake of evidence, it is necessary to distinguish among *coordinate systems*, *reference systems*, and *reference frames* [7].

A *coordinate system* is a set of rules that state the correspondence between coordinates and points; a *coordinate* is one of a set of N numbers individuating the location of a point in an N-dimensional space. A coordinate system is defined once a point known as *origin*, a set of N lines, called *axes*, all passing for the origin and having well-known relationships to each other, and a *unit length* are established.

In GPS applications, the position of a point in a coordinate system can be expressed in (Figure 2.1):

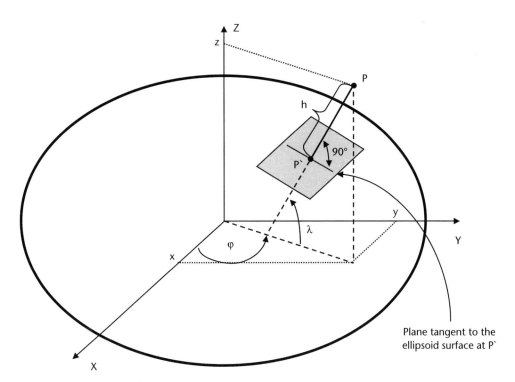

Figure 2.1   Cartesian and ellipsoidal or geodetic coordinates.

- *Cartesian coordinates (x, y, z)*;
- *Ellipsoidal* or *geodetic* (also called *geographic* [3]) *coordinates* (λ, φ, h): λ is the *latitude*, φ is the *longitude*, and h is the height above the surface of the Earth [6].

A *reference system* is the conceptual idea of a particular coordinate system. A *reference frame* is the practical realization of a reference system by observations and measurements (affected by errors), which means that a reference frame is a list of coordinates and velocities of stations (related to tectonic plate motion) placed in the area of interest, together with the estimated level of error in those values.

GPS employs three coordinate systems to answer three different demands [5]: the determination of the satellite orbits, which requires an *Earth-centered inertial* coordinate system, the estimate of the receiver position coordinates, which is better performed in an *Earth-centered earth-fixed* coordinate system, and the presentation of these in user-friendly form (i.e., in the usual geographic coordinates of latitude, longitude, and height).

Transformations from an inertial to an earth-fixed coordinate system can be found in [8].

2.1.1   Earth-Centered Inertial Coordinate System

Earth-centered inertial (ECI) *coordinate system* is often used to predict the position of earth-orbiting artificial satellites. The reason is that ECI is fixed in space relative to stars and therefore inertial. ECI is often defined as a Cartesian coordinate system, where the position (coordinates) is given as the distance from the origin along the three orthogonal axes.

The XY-plane coincides with the earth equatorial plane; the X-axis is directed from the earth center to a particular direction relative to the celestial sphere, hence being fixed; the Z-axis is normal to the XY-plane in the direction of the North Pole.

However, the irregularities of the earth motion that consist of the slow variation in orientation due to nutation and precession make the ECI system not really inertial, because the X-axis is fixed with respect to the earth motion, whereas the Z-axis moves following the equatorial plane motion due to the earth motion.

For this reason, the J2000 ECI coordinate system has been defined choosing the orientation of the axes relative to the orientation of the equatorial plane at a particular instant in time: January 1, 2000, at 12:00:00.00 UTC(USNO), where UTC stands for universal time coordinated and USNO for United States Naval Observatory. Therefore, the definition of the J2000 ECI is (Figure 2.2):

- *Origin:* Earth center of mass;
- *X-axis:* The direction of the vernal equinox at January 1, 2000, at 12:00:00.00 UTC (USNO);
- *Z-axis:* Earth rotational axis in the direction of the North Pole at January 1, 2000, at 12:00:00.00 UTC (USNO);
- *Y-axis:* Completes a right-handed, Earth-centered, orthogonal coordinate system.

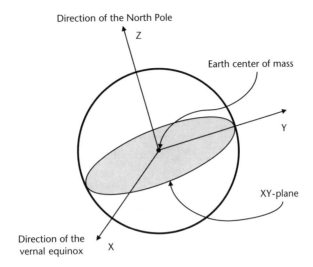

Figure 2.2   Earth-centered inertial (ECI) coordinate system.

Once defined in this way, the ECI coordinate system can be viewed as inertial [5].

2.1.2   Earth-Centered Earth-Fixed Coordinate System

The description of a receiver position in an ECI coordinate system is not user-friendly; therefore, the use of a coordinate system rotating with the earth has been introduced and indicated as *Earth-centered Earth-fixed* (ECEF) *coordinate system*: it is not inertial and can be used to define three-dimensional position in Cartesian coordinates that can then be easily transformed in latitude, longitude, and height [9–13].

In the ECEF coordinate system the origin is the earth center of mass and the XY-plane coincides with the earth equatorial plane as well as in the ECI one. However, the definition of the axes is different (Figure 2.3):

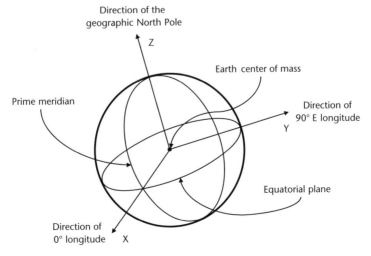

Figure 2.3   Earth-centered Earth-fixed (ECEF) coordinate system.

- *X-axis:* the direction of the line individuated by the origin and the intersection of the plane defined by the prime meridian and the equatorial plane (i.e., the direction of 0° longitude)
- *Z-axis:* the direction that points from the origin toward the geographical North Pole (i.e., the point where the meridians meet in the northern hemisphere)
- *Y-axis:* completes a right-handed, earth-centered, orthogonal coordinate system (i.e., it points to 90° East longitude).

The ECEF coordinate system is a Cartesian one, whereas GPS receivers display their own position in terms of latitude, longitude, and height, which are referred to a physical model of the Earth.

2.1.3   World Geodetic System – 1984 (WGS-84)

World Geodetic System – 1984 (WGS-84) is an ECEF reference frame and also provides a comprehensive model of the earth and information about its gravitational irregularities to compute satellite ephemeris [5]. WGS-84 can be considered as the best global geodetic reference system for the earth for different applications, such as mapping, positioning, or navigation [14].

Before giving technical and historical details about WGS-84, terminology must be introduced for a better comprehension of the upcoming issues.

2.1.3.1   Conventional Terrestrial Reference System and Frame

A *terrestrial reference system* (TRS) is a spatial reference system in which positions of points on the solid surface of the earth have coordinates undergoing only small variations in time due to tectonic or tidal deformations. In other words, a TRS rotates together with the Earth [15].

A *terrestrial reference frame* (TRF) is a collection of physical points having accurately determined coordinates in a given coordinate system, either Cartesian or geographic, fastened to a TRS [15]. Therefore, a TRF is a realization of a TRS (see also [16]), as previously specified for a generic reference frame and reference system.

When the collection of conventions, algorithms, and constants providing the origin, scale, and orientation of a TRS and their time evolution is defined, such a system is named *conventional TRS* (CTRS) [15].

The International Earth Rotation and Reference Systems Service (IERS) has the tasks of definition, realization, and diffusion of the *International Terrestrial Reference System* (ITRS), according to the International Union of Geodesy and Geophysics (IUGG) Resolution No. 2 approved in Vienna in 1991 [17]. The ITRS Product Center (ITRS-PC) provides realizations of the ITRS with the name of *International Terrestrial Reference Frame* (ITRF), whose production started in 1984 with the TRF called Bureau International de l'Heure (BIH) Terrestrial System 1984 (BTS84), realized using station coordinates obtained from different types of techniques or technologies, such as very long baseline interferometry (VLBI), lunar laser ranging (LLR), satellite laser ranging (SLR), and Doppler/TRANSIT. The BTS realizations ended with the BTS87, because, in 1988, the IUGG and the International Astronomical Union (IAU) founded the IERS. At the time of writing, the current reference realization of the ITRS is the *ITRF2000*, which has been established using station coordinates derived from VLBI, LLR, SLR, GPS, and Doppler Orbitography Radiopositioning Integrated by Satellite (DORIS), in addition to observations achieved by regional GPS networks in Alaska, North and South America, Europe, Asia, Antarctica, and the Pacific [18, 19]. The coordinates of the about 800 stations relate to the epoch 2000.0 and can be used to achieve station coordinates at another epoch through given equations and the values of velocities of the stations themselves.

2.1.3.2   WGS-84: Coordinate System Definition and Historical Background

WGS-84 is a CTRS, whose coordinate system is defined according to the IERS Technical Note 21 [20], which indicates the following criteria [14]:

- The coordinate system is geocentric and the center of mass is determined considering also oceans and atmosphere.
- The scale of the coordinate system is that of the local earth frame, in the meaning of a relativistic theory of gravitation.
- Its orientation was initially given by the BIH orientation of 1984.0.
- Its time evolution in orientation will create no residual global rotation with respect to the crust.

As shown in Figure 2.4, the origin and axes of WGS-84 coordinate system are [14]:

- *Origin:* Earth center of mass;

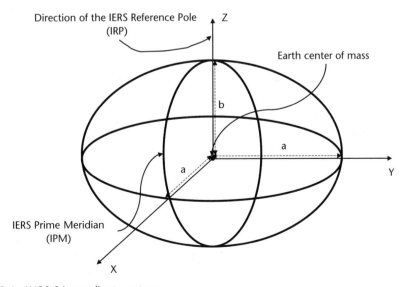

Figure 2.4   WGS-84 coordinate system.

- *Z-axis:* The direction of the IERS reference pole (IRP), which corresponds to the direction of the BIH conventional terrestrial pole (CTP) at the epoch 1984.0 with an uncertainty of 0.005″ [20];
- *X-axis:* Intersection of the IERS reference meridian (IRM, coincident with the BIH zero meridian at the epoch 1984.0 with an uncertainty of 0.005″) and the plane passing through the origin and normal to the Z-axis [20];
- *Y-axis:* Completes a right-handed, ECEF orthogonal coordinate system.

Once given the definition of the coordinate system, it is necessary to realize practically a global geodetic reference frame, as explained in the preceding section. The original WGS-84 reference frame was created in 1987 using a set of Navy Navigation Satellite System (NNSS) or TRANSIT (Doppler) station coordinates and modifying the DoD reference frame available in the early 1980s, the NSWC 9Z-2, to obtain a reference frame aligned, as closely as possible, to the BIH terrestrial system at the epoch 1984.0 [14, 21].

This TRANSIT-realized WGS-84 reference frame was adopted since January 1987 to compute the Defense Mapping Agency (DMA) TRANSIT ephemeris, which were used, in turn, to obtain the WGS-84 positions of the permanent DoD GPS monitor stations (see Chapter 3). Successively, revised sets of station coordinates were collected in two different processes, in 1994 and in 1996 through cooperation among many agencies, including the DMA (now, National Imagery and Mapping Agency, or NIMA), the Naval Surface Warfare Center Dahlgren Division (NSWCDD), and the IGS, and involving other stations besides the DoD GPS monitor stations. These refinement works resulted in two different sets of coordinates (hence two TRFs), the WGS-84 (G730) and the WGS-84 (G873). The letter "G" stands for GPS, because the coordinate estimation process was conducted, in both cases, using GPS techniques; the number following the "G" specifies the GPS week number when the related coordinates were employed in the NIMA ephemeris computation [14, 22–24].

The current realization of the WGS-84 TRF is designated as *WGS-84 (G1150)*, whose coordinates had been computed for six U.S. Air Force (USAF), 11 NIMA Monitor Station Network (MSN), two NIMA MSN development and test, three NIMA Differential GPS Reference Station (DGRS), two NSWCDD, and two IGS sites [25]. WGS-84 (G1150) is the first TRF that includes velocity estimates for each of the coordinate components. The estimated accuracy of the coordinates in WGS-84 (G1150) is 1 cm in each component.

NIMA and the GPS operational control segment (see Chapter 3) both employed WGS-84 (G1150) on January 20, 2002.

More details about the past and future evolution of WGS-84 are available in [26].

2.1.3.3   WGS-84 Ellipsoid

The original WGS-84 Development Committee selected, as geometrical reference surface approximating the Earth, a geocentric ellipsoid of revolution (Figure 2.4) to follow the approach chosen by the IUGG in the definition of the geodetic reference system 1980 (GRS 80). The parameters defining the first WGS-84 ellipsoid were the semimajor axis a, the Earth gravitational constant GM, the normalized second degree zonal gravitational coefficient $\overline{C}_{2,0}$ and the earth angular velocity ω. After successive refinements [14], it has been established that the WGS-84 ellipsoid is defined by the four parameters listed in Table 2.1.

The value of GM reported in Table 2.1 is different from the one shown in a similar table in [14], because the Aerospace Corporation suggested retaining the original GM value. The normalized second degree zonal gravitational coefficient $\overline{C}_{2,0}$ is now considered a derived geometric constant, because of the introduction, among the four defining parameters, of the flattening f.

Other derived geometric constants are:

- The semiminor axis, b, which is as long as the polar radius of the Earth, namely, 6,356,752.3142m, and is related to the flattening by the equation:

$$b = a(1 - f) \tag{2.3}$$

- The first eccentricity e given by:

$$e = \sqrt{1 - \frac{b^2}{a^2}} = 8.1819190842622 \times 10^{-2} \tag{2.4}$$

For a detailed list of all parameters and constants characterizing the WGS-84 ellipsoid, the reader has to refer to [14].

2.1.3.4   WGS-84 Relationships with Other Geodetic Systems

The WGS-84 has undergone little changes from its birth; it has always intended to be as closely coincident as possible with the CTRS adopted by the IERS or, before 1988, by the BIH. WGS-84 frame is consistent with the ITRF; for mapping and charting applications, they can be considered the same, being the differences between the two reference frames in the centimeter range worldwide.

In the European Civil Aviation Conference (ECAC) area, EUROCONTROL managed the implementation of WGS-84 in the ambit of the European Air Traffic Control Harmonization and Integration Programme (EATCHIP) to answer the request of the ICAO Council, which, in March 1989, accepted a recommendation from its Special Committee on Future Air Navigation Systems (FANS/4) [27] that stated:

Table 2.1   WGS-84 Four Defining Parameters

Parameter	Name	WGS 84 Value
Semimajor axis	a	6,378,137m
Inverse of flattening	$1/f$	298.257223563
Angular velocity	ω	7.292115×10^{-5} rad s^{-1}
Geocentric gravitational constant (mass of the Earth's atmosphere included)	GM	398600.5 km^3 s^{-2}

Recommendation 3.2/1—Adoption of WGS 84
That ICAO adopts, as a standard, the geodetic reference WGS 84 and develops appropriate ICAO material, particularly in respect to Annexes 4 and 15, in order to ensure a rapid and comprehensive implementation of the WGS 84 system.

Guidance material prepared by ICAO for WGS-84 can be found in [28].

EUREF is an International Association of Geodesy (IAG) Subcommission founded in 1987 that deals with the definition and realization of the European reference frame, based on the *European terrestrial reference system 1989* (ETRS89), whose maintenance is performed by the EUREF permanent network (EPN), composed of stations equipped with GPS/GLONASS receivers. As stated by the Resolution 1 adopted in Florence, Italy, in 1990, the ETRS89 is coincident with ITRS89; its realization can be accomplished in two different ways [29]:

- From ITRFyy (yy indicates the year in which the coordinates for the reference frame definition were collected), a correspondent ETRFyy can be derived [30]; it is presently available the reference frame *ETRF2000* [30].
- Through positioning with GPS measurements of a campaign or permanent stations [30].

2.2   Time Systems

The choice of a ranging technique such as TOA requires that satellite navigation systems have a global time reference. GPS uses the so-called *GPS system time*, which is referenced to UTC(USNO), the version of UTC maintained, at the date of July 14, 2004 [31], by a set of 59 atomic standards (including 10 hydrogen masers and 49 HP-5071 cesium atomic clocks) and using astronomical data.

UTC is a composite time scale because it is formed by inputs from two different time scales, one based on atomic clocks, the *international atomic time* (TAI), and the other based on the earth rotation rate, called *universal time 1* (UT1).

TAI is a uniform time scale defined on the atomic second, the fundamental unit of time of the international system of units (SI), and is under the responsibility of the Bureau International des Poids et Mesures (BIPM).

UT1 belongs to a family of time scales (together with UT0 and UT2), the universal time (UT), that relies on the Earth rotation on its axis; it is used to define the orientation of an ECEF coordinate system with respect to the celestial sphere and is considered the fundamental time scale for navigation [32]. Nonetheless, it is not uniform because of the variations of the earth rotation period. UT1 is maintained by the IERS.

UTC has the rate of TAI while its epoch is set to astronomic time; therefore, it was necessary to introduce *leap seconds* to maintain the difference UT1-UTC below 0.9 second in absolute value. The introduction of leap seconds is decided by the IERS and they are usually added or deleted on June 30 or December 31; the first leap second was implemented on June 30, 1972.

GPS system time is a composite time scale as UTC; its composite clock (CC) or "paper" clock comprises atomic clocks both at the monitor stations (see Chapter 3) and at the GPS satellites. GPS system time is obtained statistically processing time information coming from all these atomic clocks. Furthermore, GPS system time is referenced to the master clock (MC) at the USNO in order to not deviate by more than 1 μs from the UTC (USNO). GPS system time and UTC(USNO) were coincident on January 6, 1980.

GPS system time scale is continuous and a particular point of the line time, indicated as epoch, is identified by the number of seconds passed since Saturday/Sunday midnight and by the GPS week number, whose enumeration started with week 0 at January 6, 1980.

There is a great difference between GPS system time and UTC(USNO), which is the fact that in GPS leap seconds are not implemented; therefore, UTC(USNO) can be achieved by adding to the GPS system time the number of leap seconds introduced into UTC(USNO) [4, 5, 33].

2.3   Navigation Equations

2.3.1   Land Survey Techniques: Triangulation, Trilateration, and Traversing

Triangulation, later combined with trilateration and traversing, was used in the nineteenth and early twentieth centuries to map entire continents.

Triangulation is a land survey method based on dividing the survey area into triangles with a survey station at every corner of every triangle (Figure 2.5). The hypothesis is to know the length of one side of a triangle (for example, $\overline{AB}$ in Figure 2.5), the latitude and longitude of the points at each end of the side (A and B) and the azimuth (i.e., direction) of the side; the quantities to be measured are the angles from the points at both ends of the side to the third corner of the triangle ($B\hat{A}C$ and $A\hat{B}C$); now, the length of the other two sides of the triangle ($\overline{AC}$ and $\overline{BC}$) and the third angle ($A\hat{C}B$) can be computed, hence achieving also the latitude and longitude of the point at the third corner (C) of the triangle and the azimuth of the two sides (AC and BC). These operations are then executed for all the remaining survey stations (D and E in Figure 2.5) considering the other triangles (BCD and CDE in Figure 2.4) to obtain the coordinates of the stations and the length and the direction of the segments connecting them [34].

Also in *trilateration* the survey area is divided in triangles, but, unlike triangulation, lengths of sides of the triangles are measured instead of angles (Figure 2.6). The hypotheses are the same as those in triangulation (length $\overline{AB}$ known, latitude

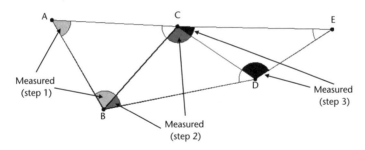

Figure 2.5   Triangulation.

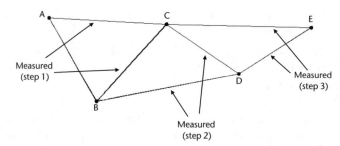

Figure 2.6   Trilateration.

and longitude of *A* and *B* known, azimuth of $\overline{AB}$ known). Measuring the length of the other two sides of the triangle ($\overline{AC}$ and $\overline{BC}$) the latitude and longitude of the point at the third corner of the triangle (*C*) can be calculated, together with the azimuth of the two sides (*AC* and *BC*). Extending these measurements and computations to all other triangles, the latitude and longitude of all other survey stations and the length and azimuth of all other segments connecting the survey stations can be obtained (Figure 2.6) [34].

Traversing uses both angle and length measurements (Figure 2.7). Supposing to know the latitude and longitude of a point (for example, *A*) and the azimuth of a segment between this point and another one (segment *AB*), the length of the segment from the point of known position (*A*) to the point of another survey station (*C*) is measured together with the angle formed by the two lines ($B\hat{A}C$). Therefore, the latitude and longitude of that point and the length and azimuth of this traverse side (*AC*) can be computed. Repeating the procedure considering all the other survey stations leads to knowing their geographic coordinates and the length and azimuth of all the segments connecting them (Figure 2.7) [34].

2.3.2   Position Determination Through Satellite Ranging Signals

The trilateration concept is used in the satellite navigation systems to fix the user position in the three-dimensional space. A *single measurement* of distance from a known point defines a *sphere of uncertainty*, as shown in Figure 2.8.

As previously discussed, range measurements are obtained through the measurement of the electromagnetic wave propagation time. The satellite position is, instead, a known figure to the user, as it is part of the received navigation message.

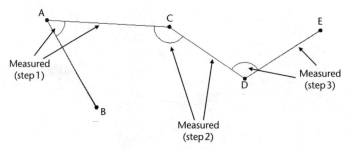

Figure 2.7   Traversing.

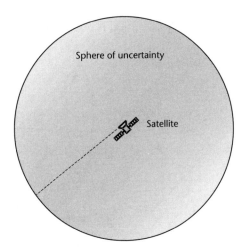

Figure 2.8   Position determination from a single satellite.

A *second range measurement*, being performed simultaneously to the first one from a further known point (i.e., satellite), allows the identification of the user position over a *circular uncertainty line* given by the intersection of the spheres centered at the satellite positions, as shown in Figure 2.9.

An additional simultaneous measure from a *third known point* (i.e., satellite) brings to intersect its uncertainty sphere with the previously mentioned circular line, allowing one to determine *two possible positions* where the user could be located Those two candidate user locations are equidistant from the plane that includes satellite positions. Therefore, the incorrect one of the two solutions can be discarded easily, as it would locate the user very far from the Earth's surface (Figure 2.10)!

Any uncertainty in the range measurement results in an uncertainty in the position estimation, as shown in Figure 2.11.

2.3.3   Basic Navigation Algorithms

This chapter's introduction lists various causes of error in the measurement of the distance between the user receiver and a satellite; to illustrate the mathematical equations at the basis of user position determination, only errors due to non-

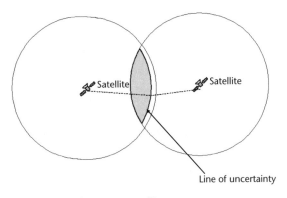

Figure 2.9   Position determination from two satellites.

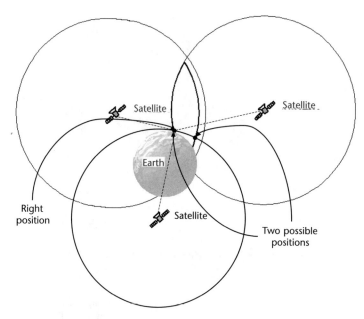

Figure 2.10   Trilateration by three satellites.

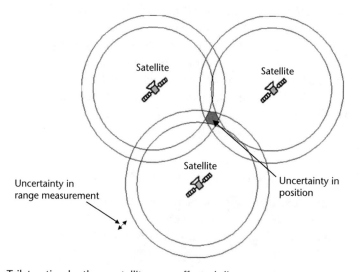

Figure 2.11   Trilateration by three satellite error-affected distance measurement.

synchronization between GPS system time, satellite clocks, and receiver clocks are taken into account.

Consider the configuration depicted in Figure 2.12, where vector $\underline{s}$ is known from the navigation message, which includes the satellite ephemeris, $\underline{r}$ is the distance vector evaluated by multiplying the TOA and the speed of light c, and $\underline{u}$ is the unknown vector to be determined.

What follows is referred to as the ECEF coordinate system, assuming that satellite coordinates are available in that format. Moving from the geometry of the problem:

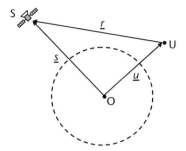

Figure 2.12   User-satellite configuration.

$$r = \|\underline{r}\| = \|\underline{s} - \underline{u}\| \tag{2.5}$$

If there were synchronization between GPS system time, satellite clock, and receiver clock, the TOA measured at the receiver would be equal to the true time interval taken by the navigation signal to travel from the satellite to the receiver. However, there is not synchronization and what is measured by the receiver is a time interval containing both the offset between GPS system time and satellite clock and the offset between GPS system time and receiver clock, as shown in Figure 2.13, where:

- t_{sat} is the GPS system time instant at which the navigation signal was broadcast from the satellite.
- t_{rec} is the GPS system time instant at which the navigation signal arrived at the user receiver.
- Δt_{sat} is the offset between GPS system time and satellite clock.
- Δt_{rec} is the offset between GPS system time and user receiver clock.

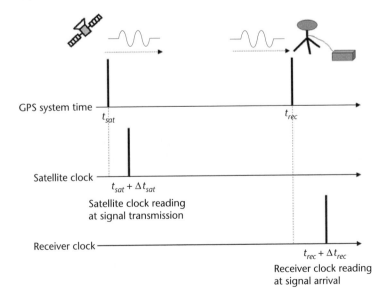

Figure 2.13   Timing relationships in the range measurement.

Therefore, satellite clock reading at the time instant of navigation signal transmission is given by $t_{sat} + \Delta t_{sat}$, whereas the receiver clock reading at the time instant of navigation signal arrival is given by $t_{rec} + \Delta t_{rec}$ [5].

The *true* or *geometric distance* between satellite and receiver is

$$r = c(t_{rec} - t_{sat}) \tag{2.6}$$

The *measured distance* between satellite and receiver (i.e., the pseudorange ρ) is

$$\rho = c\left[(t_{rec} + \Delta t_{rec}) - (t_{sat} + \Delta t_{sat})\right] = c(t_{rec} - t_{sat}) + c(\Delta t_{rec} - \Delta t_{sat}) = r + c(\Delta t_{rec} - \Delta t_{sat}) \tag{2.7}$$

The offset between GPS system time and satellite clock Δt_{sat} is compensated through corrections uploaded by the GPS control segment to the satellite, which, in turn, retransmits them to the user receiver to synchronize the transmission of the navigation signal to GPS system time. For this reason, the term Δt_{sat} will be left out in the following treatment [5].

Consequently, terms in (2.7) can be rearranged to obtain:

$$\rho = r + c\Delta t_{rec} = \|s - u\| + c\Delta t_{rec} \tag{2.8}$$

Now, let us consider the case of a user receiver undertaking TOA measurements for each satellite in visibility. In the previous section, it was shown that three measurements of distance between as many satellites and a user receiver are sufficient to obtain the receiver position, after having discarded the other point for practical reasons (Figure 2.10). This is valid in the case of coincidence between measured distances and true distances. Actually, as indicated by (2.8), the user receiver obtains a pseudorange measurement ρ; a set of three such measurements does not allow one to determine the user receiver position (Figure 2.11). Therefore, it is clear that more than three distance measurements are necessary.

This conclusion can be reached also by reasoning mathematically.

Relating (2.8) to a given ith satellite and expanding it, the following equation is achieved:

$$\rho_i = \sqrt{(x_i - x_u)^2 + (y_i - y_u)^2 + (z_i - z_u)^2} + c\Delta t_{rec} \tag{2.9}$$

where (x_i, y_i, z_i) and (x_u, y_u, z_u) denote the three-dimensional ith satellite and user position, respectively. The unknowns of the problem, hence, are *four*: x_u, y_u, z_u, and Δt_{rec}. As a consequence, at least four independent equations are necessary for solving the system:

$$\begin{aligned}
\rho_1 &= \sqrt{(x_1 - x_u)^2 + (y_1 - y_u)^2 + (z_1 - z_u)^2} + c\Delta t_{rec} \\
\rho_2 &= \sqrt{(x_2 - x_u)^2 + (y_2 - y_u)^2 + (z_2 - z_u)^2} + c\Delta t_{rec} \\
\rho_3 &= \sqrt{(x_3 - x_u)^2 + (y_3 - y_u)^2 + (z_3 - z_u)^2} + c\Delta t_{rec} \\
\rho_4 &= \sqrt{(x_4 - x_u)^2 + (y_4 - y_u)^2 + (z_4 - z_u)^2} + c\Delta t_{rec}
\end{aligned} \tag{2.10}$$

This nonlinear equation system can be solved by various techniques: *closed form solutions* [35–40] or *iterative solutions* based on either *linearization* [41–45] or on *Kalman filtering* [46–50]. A detailed description of the filtering and state estimation algorithms for the position determination is beyond the scope of this book and can be found in [51–57]. However, it is important here to render the reader somehow aware of a calculation algorithm that is implemented in many commercial receivers [6, 58], in order to provide a clear connection between the theoretical basis of navigation and its translation into practical implementation. In what follows, the *linearization approach* for (2.10) and the *least squares estimation* that bring to the mentioned algorithms are detailed. The receiver clock error will be multiplied by c, hence becoming a length (in meters), denoted by $b = c\Delta t_{rec}$.

The iterative scheme is based on the knowledge of an estimate of the user position and the clock offset that will be indicated by $(\hat{x}_u, \hat{y}_u, \hat{z}_u, \hat{b})$. The true position (and the true clock offset) is the sum of the estimate position (and the estimate clock offset) and a displacement $(\Delta x_u, \Delta y_u, \Delta z_u, \Delta b)$:

$$
\begin{aligned}
x_u &= \hat{x}_u + \Delta x_u \\
y_u &= \hat{y}_u + \Delta x_u \\
z_u &= \hat{z}_u + \Delta x_u \\
b &= \hat{b} + \Delta b
\end{aligned}
\tag{2.11}
$$

An approximate pseudorange can be also defined as follows:

$$
\hat{\rho}_i = \sqrt{(x_i - \hat{x}_u)^2 + (y_i - \hat{y}_u)^2 + (z_i - \hat{z}_u)^2} + \hat{b} = f\left(\hat{x}_u, \hat{y}_u, \hat{z}_u, \hat{b}\right)
\tag{2.12}
$$
$$
i = 1, 2, 3, 4
$$

As a consequence, the unknowns of the problem become $(\Delta x_u, \Delta y_u, \Delta z_u, \Delta b)$; in fact:

$$
\rho_i = f(x_u, y_u, z_u, b) = f\left(x_u + \Delta x_u, \hat{y}_u + \Delta y_u, \hat{z}_u + \Delta z_u, \hat{b} + \Delta b\right)
\tag{2.13}
$$
$$
i = 1, 2, 3, 4
$$

The latter function can be expanded with respect to the approximate point and estimated receiver clock offset using a Taylor series truncated after first order:

$$
f\left(\hat{x}_u + \Delta x_u, \hat{y}_u + \Delta y_u, \hat{z}_u + \Delta z_u, \hat{b} + \Delta b\right) \cong f\left(\hat{x}_u, \hat{y}_u, \hat{z}_u, \hat{b}\right) + \frac{\partial f\left(\hat{x}_u, \hat{y}_u, \hat{z}_u, \hat{b}\right)}{\partial \hat{x}_u}\Delta x_u
$$
$$
+ \frac{\partial f\left(\hat{x}_u, \hat{y}_u, \hat{z}_u, \hat{b}\right)}{\partial \hat{y}_u}\Delta y_u + \frac{\partial f\left(\hat{x}_u, \hat{y}_u, \hat{z}_u, \hat{b}\right)}{\partial \hat{z}_u}\Delta z_u + \frac{\partial f\left(\hat{x}_u, \hat{y}_u, \hat{z}_u, \hat{b}\right)}{\partial \hat{b}}\Delta b
$$
$$
\tag{2.14}
$$

The partial derivatives can be evaluated using (2.12) as follows:

$$\frac{\partial f\left(\hat{x}_u, \hat{y}_u, \hat{z}_u, \hat{b}\right)}{\partial \hat{x}_u} = -\frac{x_i - \hat{x}_u}{\hat{D}_i}$$

$$\frac{\partial f\left(\hat{x}_u, \hat{y}_u, \hat{z}_u, \hat{b}\right)}{\partial \hat{y}_u} = -\frac{y_i - \hat{y}_u}{\hat{D}_i}$$

$$\frac{\partial f\left(\hat{x}_u, \hat{y}_u, \hat{z}_u, \hat{b}\right)}{\partial \hat{z}_u} = -\frac{z_i - \hat{z}_u}{\hat{D}_i} \tag{2.15}$$

$$\frac{\partial f\left(\hat{x}_u, \hat{y}_u, \hat{z}_u, \hat{b}\right)}{\partial \hat{b}} = 1$$

where:

$$\hat{D}_i = \sqrt{\left(x_i - \hat{x}_u\right)^2 + \left(y_i - \hat{y}_u\right)^2 + \left(z_i - \hat{z}_u\right)^2} \quad i = 1, 2, 3, 4 \tag{2.16}$$

indicates the estimation of the geometrical distance between satellites and the receiver estimated position. The first three derivatives in (2.15) indicate the *direction cosines* of the unit vector pointing from the ith satellite to the estimated user terminal position.

By substituting (2.15), (2.12), and (2.13) into (2.14), we have:

$$\rho_i = \hat{\rho}_i - \frac{x_i - \hat{x}_u}{\hat{D}_i} \Delta x_u - \frac{y_i - \hat{y}_u}{\hat{D}_i} \Delta y_u - \frac{z_i - \hat{z}_u}{\hat{D}_i} \Delta z_u + \Delta b \tag{2.17}$$

By indicating with l_i, m_i, n_i, the estimated direction cosines referred to the ith satellite, (2.17) can be rewritten as follows:

$$\hat{\rho}_i - \rho_i = \Delta \rho_i = l_i \Delta x_u + m_i \Delta y_u + n_i \Delta z_u - \Delta b$$
$$i = 1, 2, 3, 4 \tag{2.18}$$

or in matrix form:

$$\underline{\Delta \rho} = H \underline{\Delta x} \tag{2.19}$$

which has solution:

$$\underline{\Delta x} = H^{-1} \underline{\Delta \rho} \tag{2.20}$$

where:

$$\underline{\Delta \rho} = \begin{bmatrix} \Delta \rho_1 \\ \Delta \rho_2 \\ \Delta \rho_3 \\ \Delta \rho_4 \end{bmatrix}; H = \begin{bmatrix} l_1 & m_1 & n_1 & 1 \\ l_2 & m_2 & n_2 & 1 \\ l_3 & m_3 & n_3 & 1 \\ l_4 & m_4 & n_4 & 1 \end{bmatrix}; \underline{\Delta x} = \begin{bmatrix} \Delta x_u \\ \Delta y_u \\ \Delta z_u \\ -\Delta b \end{bmatrix} \tag{2.21}$$

This formulation of the problem brings to an iterative resolution scheme, where—moving from a given approximate solution—corrections $\underline{\Delta x}$ are calculated to create the basis for a new computation that improves the previous approximation.

Although four simultaneous pseudorange measurements from four different satellites are adequate to achieve a solution, the iterative scheme could bring an algorithm divergence in some cases, due to both an erroneous choice of the initial estimate and to systematic errors that might be compensated only through several iterations.

These reasons bring an *overdetermined pseudorange calculation approach,* using more than four satellites.

The described formulation generally results to be inconsistent, as small errors in the pseudorange measurement prevent identifying a single $\underline{\Delta x}$ value able to solve the system exactly.

The least squares method can be, hence, used to achieve the position estimation and the user clock bias.

The following methodology is very general and can be exploited also in problems different from satellite navigation.

If the satellite number N is higher than four, the system (2.19) has more equations than unknowns, with H an $N \times 4$ matrix and $\underline{\Delta \rho}$ a column vector $N \times 1$.

By defining a vector quantity $\underline{v}$, denominated *residual*:

$$\underline{v} = H\underline{\Delta x} - \underline{\Delta \rho} \tag{2.22}$$

a value of $\underline{\Delta x}$ is searched so that $\underline{\Delta \rho}$ be very close to $H\underline{\Delta x}$; in case $\underline{\Delta \rho} = H\underline{\Delta x}$, the residual (2.22) is zero.

The least squares solution is defined as the $\underline{\Delta x}$ value that minimizes the square of the residual $\underline{v}$, or, equivalently, the sum of the squares of the $\underline{v}$ components.

The residual $\underline{v}$ is, hence, a function of $\underline{\Delta x}$. In the following, we indicate with v_q the square of $\underline{v}$:

$$v_q(\underline{\Delta x}) = \left(H\underline{\Delta x} - \underline{\Delta \rho}\right)^2 = \left(H\underline{\Delta x} - \underline{\Delta \rho}\right)^T \left(H\underline{\Delta x} - \underline{\Delta \rho}\right) \tag{2.23}$$

By differentiating (2.23) with respect to $\underline{\Delta x}$ we obtain the gradient vector $\underline{\nabla v_q}$ of v_q. The minimum of v_q is achieved for the $\underline{\Delta x}$ value that brings to zero $\underline{\nabla v_q}$. This is, in principle, only a necessary condition; however, it can be demonstrated that in the least squares problem the zero-gradient-condition is also sufficient to minimize v_q, because the v_q Hessian is nonnegative. Accounting for:

$$\left(\underline{\Delta \rho}\right)^T H\underline{\Delta x} = \left(\underline{\Delta x}\right)^T H^T \underline{\Delta \rho} \tag{2.24}$$

(2.23) can be rewritten as:

$$v_q(\underline{\Delta x}) = \left(\underline{\Delta x}\right)^T H^T H\underline{\Delta x} - 2\left(\underline{\Delta x}\right)^T H^T \underline{\Delta \rho} + \left\|\underline{\Delta \rho}\right\|^2 \tag{2.25}$$

The gradient of v_q, that is, a function of four unknowns, can be calculated referring to two simple relationships of linear algebra. One refers to the scalar $\underline{y}^T\underline{x}$ or $\underline{x}^T\underline{y}$:

$$\frac{\partial}{\partial \underline{x}} \left(\underline{y}^T \underline{x} \right) = \frac{\partial}{\partial \underline{x}} \left(\underline{y}^T \underline{x} \right) = \underline{y} \qquad (2.26)$$

The second relationship refers to square forms (i.e., to the scalars that can be rewritten as $\underline{x}^T A \underline{x}$):

$$\frac{\partial}{\partial \underline{x}} \left(\underline{x}^T A \underline{x} \right) = 2A^T \underline{x} = 2A\underline{x} \qquad (2.27)$$

Equation (2.27) holds true in case the A matrix is symmetric; in our case the role of A of the (2.27) is taken by $H^T H$, clearly symmetric.

By differentiating (2.25) and exploiting the properties (2.26) and (2.27) we obtain:

$$\nabla v_q = 2H^T H \underline{\Delta x} - 2H^T \underline{\Delta \rho} \qquad (2.28)$$

By imposing that (2.28) is zero (i.e., the four components of the 4×1 vector $\underline{\nabla v_q}$ be zero), the solution to the system in the case $H^T H$ is nonsingular (i.e., H columns are independent and matrix H has full rank) is:

$$\underline{\Delta x} = \left(H^T H \right)^{-1} H^T \underline{\Delta \rho} \qquad (2.29)$$

Equation (2.29) is the least squares solution of the satellite navigation equations. In case we have only four pseudorange measurements, H becomes again a 4×4 matrix and (2.29) is identical to (2.20), accounting that $(H^T H)^{-1} = H^{-1} (H^T)^{-1}$.

The least squares method provides, hence, an optimum estimation of the unknown $\underline{\Delta x}$, as it minimizes the residual square together with the estimation variance.

Least square estimators are *best linear unbiased estimator* (BLUE), where *best* means they have the least variance and are, thus, efficient; *linear* means they are in the form of linear combination of dependant variables, *unbiased* means the expected value of the estimator equals the true value of the coefficient.

2.4   Required Navigation Performance (RNP)

The ICAO developed the RNP approach in the early 1990s and it is a standard for airspace requirements nowadays. RNP is a statement of the navigation performance necessary for operation within a defined airspace.

RNP concepts could be used to define requirements both for satellite navigation systems and terrestrial radio navigation systems. The strict necessities of aviation safety have brought precise definitions of almost all the characteristics of navigation parameters. This approach is suitable for nonaeronautic services as well and is used to define user requirements for several navigation projects.

Most of the RNP can be described only statistically by a probability density function and the requirements are often given by probabilistic bounds. The main required navigation parameters defined in [59] are the following:

- *Accuracy*: The degree of conformance between the estimated or measured position and/or the velocity of a platform at a given time and its true position or velocity. Radio navigation performance accuracy is usually presented as a statistical measure of system error and is specified as:
  - *Predictable*: The accuracy of a position in relation to the geographic or geodetic coordinates of the Earth.
  - *Repeatable*: The accuracy with which a user can return to a position whose coordinates have been measured at a previous time with the same navigation system.
  - *Relative*: The accuracy with which a user can determine one position relative to another position regardless of any error in their true position.
- *Integrity*: The ability of a system to provide timely warnings to users when the system should not be used for navigation. In particular, the system is required to deliver to the user an alert within the *time to alert* when an *alert limit* is exceeded. The alert limit is the maximum error allowable in the user-computed position solution; the alert limit can be specified in horizontal alert limit (HAL) and vertical alert limit (VAL).
- *Integrity risk:* The probability during the period of operation that an error, whatever the source, will result in a computed position error exceeding the alert limit, and that the user will not be informed within the specified time-to-alert.
- *Continuity*: The continuity of a system is the capability of the total system (including all elements necessary to maintain aircraft position within the defined airspace) to perform its function without nonscheduled interruptions during the intended operation. The continuity risk is the probability that the system will be unintentionally interrupted and will not provide guidance information for the intended operation. More specifically, continuity is the probability that the system will be available for the duration of a phase of operation, presuming that the system was available at the beginning of that phase of operation.
- *Availability*: The availability of a navigation system is the percentage of time that the system is performing a required function under stated conditions. Availability is an indication of the ability of the system to provide usable service within the specified coverage area. Signal availability is the percentage of time that navigation signals transmitted from external sources are available for use.

References

[1]   Tetley, L., and D. Calcutt, *Electronic Navigation Systems*, 3rd ed., Boston, MA: Butterworth-Heinemann, 2001.

[2]   Powers, E., et al., "Potential Timing Improvements in GPS III," *37th Meeting of the Civil GPS Service Interface Committee*, Arlington, VA, March 2001; http://www.navcen.uscg.gov/cgsic/meetings/summaryrpts/37thmeeting/default.htm.

[3]   El-Rabbany, A., *Introduction to GPS: The Global Positioning System*, Norwood, MA: Artech House, 2002.

[4]   Parkinson, B. W., and J. J. Spilker Jr., (eds.), "Global Positioning System: Theory and Applications," *Progress in Astronautics and Aeronautics, American Institute of Aeronautics and Astronautics*, Vols. 163 and 164, 1996.

[5]   Kaplan, E. D., *Understanding GPS: Principles and Applications*, Norwood, MA: Artech House, 1996.

[6]   Bao-Yen Tsui, J., *Fundamentals of Global Positioning System Receivers: A Software Approach*, New York: Wiley-Interscience, 2000.

[7]   *Geodetic Glossary*, National Geodetic Survey, 1996.

[8]   Long, A. C., et al., (eds.), "Goddard Trajectory Determination System (GTDS) Mathematical Theory," Revision 1, FDD/552-89/001, Goddard Space Flight Center, Greenbelt, MD, July 1989.

[9]   Gaposchkin, P., "Reference Coordinate Systems for Earth Dynamics," *Proceedings of the 56th Colloquium of the International Astronomical Union*, Warsaw, Poland, July 1981.

[10] Minkler, G., and J. Minkler, *Aerospace Coordinate Systems and Transformations*, Adelaide, Australia: Magellan Book Co., 1990.

[11] Wolper, J. S., *Understanding Mathematics for Aircraft Navigation*, New York: McGraw-Hill, 2001.

[12] Sudano, J. J., "An Exact Conversion from an Earth-Centered Coordinate System to Latitude, Longitude and Altitude," *Aerospace and Electronics Conference, 1997, NAECON 1997, Proceedings of the IEEE 1997 National*, Vol. 2, July 14–17, 1997, pp. 646–650.

[13] Maling, D. H., *Coordinate Systems and Map Projections*, 2nd ed., New York: Pergamon Press, 1992.

[14] *Department of Defense World Geodetic System 1984—Its Definition and Relationships with Local Geodetic Systems*, NIMA Technical Report TR8350.2, 3rd ed., Amendment 1, January 3, 2000, updated on June 23, 2004; http://www.earth-info.nima.mil/GandG/tr8350/tr8350_2.html.

[15] McCarthy, D. D., and G. Petit, (eds.), "IERS Conventions (2003)," *IERS Technical Note No. 32*, IERS Conventions Centre, U.S. Naval Observatory (USNO), Bureau International des Poids et Mesures (BIPM), 2004; http://www.iers.org/iers/publications/tn/tn32/.

[16] Boucher, C., "Terrestrial Coordinate Systems and Frames," *Encyclopedia of Astronomy and Astrophysics*, Version 1.0, Bristol, England: Nature Publishing Group and Institute of Physics Publishing, 2001, pp. 3289–3292.

[17] *Geodesist's Handbook*, Delft, the Netherlands: Bulletin Géodésique, Vol. 66, 1992.

[18] Boucher, C., et al., "The ITRF2000," *IERS Technical Note No. 31*, IERS ITRS Centre, Institut Géographique National (IGN), Laboratoire de Recherche en Geodesie (LAREG), Ecole Nationale de Sciences Geographiques (ENSG), 2004; http://www.iers.org/iers/publications/tn/tn31/.

[19] Altamimi, Z., P. Sillard, and C. Boucher, "ITRF2000: A New Release of the International Terrestrial Reference Frame for Earth Science Applications," *Journal of Geophysical Research*, Vol. 107, No. B10, 2002, p. 2214.

[20] McCarthy, D., (ed.), "IERS Conventions (1996)," *IERS Technical Note No. 21*, U.S. Naval Observatory, July 1996; http://www.maia.usno.navy.mil/conventions.html.

[21] *WGS 84 Implementation Manual*, Version 2.4, EUROCONTROL (European Organization for the Safety of Air Navigation, Brussels, Belgium), Institute of Geodesy and Navigation (IfEN, University FAF, Munich, Germany), February 1998.

[22] Swift, E. R., "Improved WGS 84 Coordinates for the DMA and Air Force GPS Tracking Sites," *Proc. of ION GPS-94*, Salt Lake City, UT, September 1994.

[23] Cunningham, J., and V. L. Curtis, "WGS 84 Coordinate Validation and Improvement for the NIMA and Air Force GPS Tracking Stations," NSWCDD/TR-96/201, November 1996.

[24] Malys, S., and J. A. Slater, "Maintenance and Enhancement of the World Geodetic System 1984," *Proc. of ION GPS-94*, Salt Lake City, UT, September 1994.

[25] Merrigan, M. J., et al., "A Refinement to the World Geodetic System 1984 Reference Frame," *Proc. of ION GPS-2002*, The Institute of Navigation, Portland, OR, September 2002.

[26] True, S. A., "Planning the Future of the World Geodetic System 1984," *Proc. of the Position, Location and Navigation Symposium 2004*, Monterey, CA, April 26–29, 2004.

[27] *Conventions on International Civil Aviation, Annex 15: Aeronautical Information Services*, International Civil Aviation Organization, Montreal, ICAO, 2003.

[28] *World Geodetic System – 1984 (WGS-84) Manual*, Doc. 9674, 2nd ed., International Civil Aviation Organization, 2002.

[29] EUREF Web site http://www.lareg.ensg.ign.fr/EUREF.

[30] Boucher, C., and Z. Altamimi, "Specifications for Reference Frame Fixing in the Analysis of a EUREF GPS Campaign," Version 5, 2001, http://eareg.ensg.ign.fr/EUREF/memo.ps.

[31] CGI Script at the Web site http://www.tycho.usno.navy.mil/time_scale.html.

[32] Seeber, G., *Satellite Geodesy: Foundations, Methods, and Applications*, New York: Walter De Gruyter, 1993.

[33] USNO Time Service Web site http://www.tycho.usno.navy.mil.

[34] *Geodesy for the Layman*, Defense Mapping Agency, December 1983; http://www.earth-info.nga.mil/GandG/pubs.html).

[35] Abel, J. S., and J. W. Chaffee, "Existence and Uniqueness of GPS Solutions," *IEEE Trans. on Aerospace and Electronic Systems*, Vol. 27, Issue 6, November 1991, pp. 952–956.

[36] Fang, B. T., "Comments on 'Existence and Uniqueness of GPS Solutions' by J. S. Abel and J. W. Chaffee," *IEEE Trans. on Aerospace and Electronic Systems*, Vol. 28, Issue 4, October 1992, p 1163.

[37] Phatak, M., M. Chansarkar, and S. Kohli, "Position Fix from Three GPS Satellites and Altitude: A Direct Method," *IEEE Trans. on Aerospace and Electronic Systems*, Vol.35, Issue 1, January 1999, pp. 350–354.

[38] Hoshen, J., "On the Apollonius Solutions to the GPS Equations," *AFRICON, 1999 IEEE*, Vol. 1, September 28–October 1, 1999, pp. 99–102.

[39] Leva, J. L., "An Alternative Closed-Form Solution to the GPS Pseudorange Equations," *IEEE Trans. on Aerospace and Electronic Systems*, Vol. 32, Issue 4, October 1996, pp. 1430–1439.

[40] Chaffee, J., and J. Abel, "On the Exact Solutions of Pseudorange Equations," *IEEE Trans. on Aerospace and Electronic Systems*, Vol. 30, Issue 4, October 1994, pp. 1021–1030.

[41] Hassibi, B., and H. Vikalo, "On the Expected Complexity of Integer Least-Squares Problems," *IEEE International Conference on Acoustics, Speech, and Signal Processing*, Vol. 2, 2002, pp. 1497–1500.

[42] Hassibi, A., and S. Boyd, "Integer Parameter Estimation in Linear Models with Applications to GPS," *IEEE Trans. on Signal Processing*, [see also *IEEE Trans. on Acoustics, Speech, and Signal Processing*], Vol. 46, Issue 11, November 1998, pp. 2938–2952.

[43] Abel, J. S., "A Divide and Conquer Approach to Least-Squares Estimation with Application to Range-Difference-based Localization," *International Conference on Acoustics, Speech, and Signal Processing*, Vol. 4, May 23–26, 1989, pp. 2144–2147.

[44] Peng, H. M., et al., "Maximum-Likelihood-Based Filtering for Attitude Determination Via GPS Carrier Phase," *IEEE Position Location and Navigation Symposium*, March 13–16, 2000, pp. 480–487.

[45] Hassibi, B., and H. Vikalo, "On the Expected Complexity of Sphere Decoding," *Conference Record of the Thirty-Fifth Asilomar Conference on Signals, Systems and Computers*, Vol. 2, November 4–7, 2001, pp. 1051–1055.

[46] Chaffee, J. W., and J. S. Abel, "The GPS Filtering Problem," *Position Location and Navigation Symposium*, "Record '500 Years After Columbus—Navigation Challenges of Tomorrow'," *IEEE PLANS '92*, March 23–27, 1992, pp. 12–20.

[47] Ponomaryov, V. I., et al., "Increasing the Accuracy of Differential Global Positioning System by Means of Use the Kalman Filtering Technique," *Proc. of the 2000 IEEE International Symposium on Industrial Electronics*, Vol. 2, December 4–8, 2000, pp. 637–642.

[48] Mao, X., M. Wada, and H. Hashimoto, "Investigation on Nonlinear Filtering Algorithms for GPS," *IEEE Intelligent Vehicle Symposium*, Vol. 1, June 17–21, 2002, pp. 64–70.

[49] Mao, X., M. Wada, and H. Hashimoto, "Nonlinear Filtering Algorithms for GPS Using Pseudorange and Doppler Shift Measurements," *Proc. of the 5th IEEE International Conference on Intelligent Transportation Systems*, Singapore, 2002, pp. 914–919.

[50] Wu, S.-C., and W. G. Melbourne, "An Optimal GPS Data Processing Technique for Precise Positioning," *IEEE Trans. on Geoscience and Remote Sensing*, Vo. 31 Issue 1, January 1993, pp. 146–152.

[51] Nardi, S., and M. Pachter, "GPS Estimation Algorithm Using Stochastic Modelling," *Proc. of the 37th IEEE Conference on Decision and Control*, Vol. 4, Tampa, FL: December 16–18, 1998.

[52] Chaffee, J., J. Abel, and B. K. McQuiston, "GPS Positioning, Filtering, and Integration," *Proc. of the IEEE 1993 National Aerospace and Electronics Conference*, Dayton, OH, Vol. 1, May 24–28, 1993, pp. 327–332.

[53] Zhuang, W., and J. Tranquilla, "Modeling and Analysis for the GPS Pseudorange Observable," *IEEE Trans. on Aerospace and Electronic Systems*, Vol. 31, No. 2, April 1995, pp. 739–751.

[54] Fenwick, A. J., "Algorithms for Position Fixing Using Pulse Arrival Times," *IEE Proc. on Radar, Sonar and Navigation*, Vol. 146, No. 4, August 1999, pp. 208–212.

[55] Shin, D.-H., and Tae-Kyung Sung, "Comparisons of Error Characteristics Between TOA and TDOA Positioning," *IEEE Trans. on Aerospace and Electronic Systems*, Vol. 38, No. 1, January 2002, pp. 307–311.

[56] Chaffee, J. W., "Observability, Ensemble Averaging and GPS Time," *IEEE Trans. on Aerospace and Electronic Systems*, Vol. 28, No. 1, January 1992, pp. 224–240.

[57] Abel, J. S., "A Variable Projection Method for Additive Components with Application to GPS," *IEEE Trans. on Aerospace and Electronic Systems*, Vol. 30, No. 3, July 1994, pp. 928–930.

[58] Xu, G., *GPS: Theory, Algorithms, and Applications*, New York: Springer-Verlag, 2003.

[59] "Manual on Required Navigation Performance (RNP)," *ICAO*, Doc. 9613, June 1999.

CHAPTER 3

The Global Positioning System

3.1 Introduction

In describing the evolution of satellite navigation (see Section 1.2 in Chapter 1), a contradictory and fascinating portrait of the GPS can be derived.

GPS, the system used for positioning, tracking, and mapping in most cases is mentioned as synonymous with navigation; GPS is the means that has translated the theoretical concept of navigation (see Chapter 2) into an actual system, a quite friendly receiver, a commonly accepted and increasingly needed service.

It is designed to be a low-cost solution for civilian users; however, it has been conceived and continues to be maintained by U.S. military institutions, albeit with a contribution from the U.S. Department of Transportation.

It is global in scope, available on a 24/7/365 basis (i.e., always), but it can be denied to civilian users due to defense-based security reasons.

It is accurate, but its performance could be intentionally degraded (as was the case for more than 10 years until 2000) to ensure U.S. national security by denying accurate real-time autonomous navigation to unauthorized users.

GPS has been a continuous source of technology deployment for onboard and on-ground hardware and software. It has brought into action a competitive scenario for conceiving and deploying future systems and innovative services that render navigation a key aspect of people's future welfare.

Extensive literature has been written on GPS [1–9], including very detailed books and research papers [10–12]. Hence, these will not be reproduced here. However, the GPS architecture and organization, signal structure, and performance issues are covered here. An understanding of these issues is key to comprehending the evolution and future directions of navigation systems. This chapter provides a view on these topics, along with design-related considerations, that should stimulate a personal vision of the possible developments and potential of the navigation world.

3.2 GPS System Architecture

GPS is a complex system that comprises *space*, *control*, and *user* segments. This section highlights the major features of the three segments: the aim is to delineate the translation of the basic navigation concept described in Chapter 2 into an actual and successful operative system, which has brought people into the navigation world and concurred to develop the awareness of usefulness and further potential of navigation services.

3.2.1 Space Segment

GPS uses a constellation of satellites, each transmitting a composite ranging signal that includes a navigation message [10, 12–15]. The latter contains the information required to determine the coordinates of the satellites and bring the satellite clocks in line with the GPS time. As outlined in Chapter 2, measurements are needed simultaneously from at least four satellites (Figure 3.1) to determine three-dimensional positioning and timing capability.

A number of considerations have gone into specifying the GPS satellite constellation. The satellite constellation provides ranging capability for any user located anywhere and at anytime on Earth. In addition, satellites need to be widely spaced in angle to minimize their mutual interference during the simultaneous measurement operations and to provide the required geometric strength for high-precision positioning. Finally, satellite height should guarantee a suitable compromise between the fraction of Earth visible from each satellite of the constellation and the receiver complexity related to Doppler shift, signal acquisitions, and intersatellite handovers.

As a result, the selected GPS constellation contains 24 satellites in six orbital planes (see Figure 3.2). Each plane hosts four satellites. The satellites have a period of 12 hours sidereal time, corresponding to about 11 hours and 58 minutes, and a height of about 20,163 km above the Earth. Orbits are inclined at 55° with respect to the equatorial plane and have a nearly circular shape (the eccentricity is about 0.01).

The GPS constellation became fully operational in 1995 with 24 nonexperimental satellites in proper use and has always included more than 24 active satellites and back-up units to improve operation reliability in case of nonscheduled satellite interruptions. GPS satellites weigh about 930 kg and are 5.2m long with the solar panels extended. They adopt thrusters for orbit adjustment and have a design life of 7.5 years.

The payload is designed to transmit properly in time the composite L-band signals, L1 and L2, containing the navigation data, which provide users with the necessary information to determine the positioning, velocity, and timing measurements (see Section 3.3). The navigation data are periodically (up to twice a day) updated onboard through the S-band tracking, telemetry and command (TT&C) system by on-ground control segment and stored in the satellite memory.

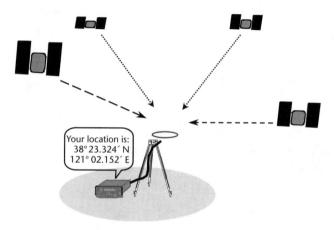

Your location is:
38° 23.324′ N
121° 02.152′ E

Figure 3.1 The GPS basic concept.

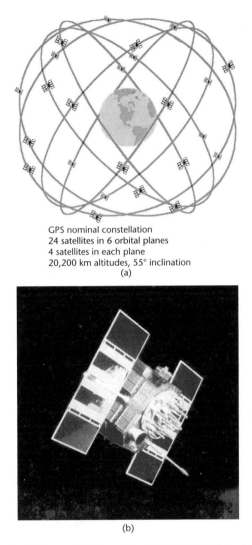

GPS nominal constellation
24 satellites in 6 orbital planes
4 satellites in each plane
20,200 km altitudes, 55° inclination
(a)

(b)

Figure 3.2 (a) The nominal GPS constellation; (b) a NAVSTAR satellite of the GPS constellation.

The core of the satellite payload are the redundant atomic oscillators that achieve the required high stability through either rubidium or cesium atoms in gaseous form. The atomic clocks, together with proper frequency synthesizers, then synchronize the GPS signal generators, also controlling the L-band center frequencies. Signals are then amplified, filtered, and modulated to create the proper format to be transmitted toward the user segment (see Section 3.3).

3.2.2 Control Segment

Monitoring of the GPS satellites, through checks of their operational health and determining their positions in space, is carried out by the operational control segment (OCS) [1, 10, 12]. In particular, the segment takes care of: maintaining the satellites in due orbit through small maneuvers; introducing corrections and adjustments to satellite clocks and payload; tracking the GPS satellites and uploading navigation data to each satellite of the constellation; and providing through commands major relocations in case of satellite failure.

The OCS, whose operations started in 1985, consists of a network of tracking stations (Figure 3.3): the master control station (MCS) is located at Falcon Air Force Base in Colorado Springs; five monitor stations (MS), whose coordinates are known with extremely high precision, are located in Hawaii, Colorado Springs (colocated with the MCS), Ascension Island, Diego Garcia in the Indian Ocean, and Kwajalein in the West Pacific. The monitor stations are equipped with high-precision GPS receivers and a cesium oscillator for continuous tracking of all visible satellites. The premises at Ascension Island, Diego Garcia, and Kwajalein are also equipped with ground antennas (GA) for uploading information to the GPS satellites through the S-band TT&C link. Monitor stations are unmanned and operated remotely from the MCS.

The MCS receives GPS observations from the MSs, collected through their multiple code and frequency GPS tracking receivers, and processes them to generate a set of predicted navigation data. Updated navigation data are sent by the MCS to a GA for uplink to the satellites. The satellites incorporate these updates in the signals that they transmit to the GPS receivers. These operations are highlighted in Figure 3.4.

As previously outlined, the OCS also has the task of monitoring the GPS system health. In particular, the status of a satellite is set to unhealthy by the MCS during satellite maintenance or outage. This information is included in the satellite navigation message, although the civilian service is not monitored for health information in real time. Constellation status and any scheduled or unscheduled event that affects

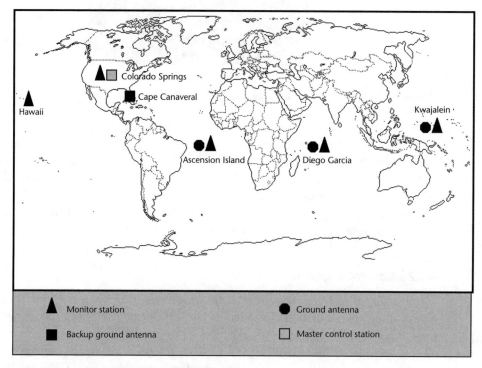

Figure 3.3   Location map of GPS OCS stations.

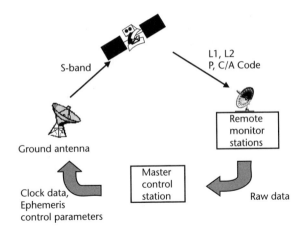

Figure 3.4 Basic structure and data flow of the GPS control segment.

the service are timely (at least 48 hours prior to the event in the first case and as soon as possible in the second one) notified by the U.S. Coast Guard Navigation Center (NAVCEN) in cooperation with the Navigation Information Service (NIS). The NAVCEN, hence, provides users with notifications on the GPS status through the Notice: Advisory to Navigation Users (NANU). This information is the primary input in the generation of GPS status report and notices, such as Notice to Airmen (NOTAM) and U.S. Coast Guard Local Notice to Mariners (LNM). In addition, constellation status information about scheduled satellite maintenance or outage can be accessed to the public through, for instance, Internet-based information [16].

From this description, the MCS is clearly the core of the control segment, as it tracks and controls other stations and, through them, the whole satellite constellation, providing the proper reliability to the system.

3.2.3 User Segment

In a complex system architecture the *user segment* is the place where very different elements, all playing a key role in the success of the system itself, converge:

- The *user terminal* (i.e., the closest-to-the-user hardware component of the system, whose cost and use-friendship helps the user determine if the system's financial and access compliance meet with his or her expectations or constraints);
- The *application* (i.e., the demonstration at software-oriented level of the understanding from the system designer and operator of the actual user needs).

User-friendly and low-cost terminals together with user-need-based applications, flexibly conceived to be able to be extended and/or modified according to the evolving scenario, are keys for system effectiveness and success. GPS certainly addresses these aspects, allowing users worldwide to access present and potential applications of navigation systems through friendly and quite cheap terminals [10–12, 17–20].

In the GPS receiver satellite signals are received via a right-hand circularly polarized antenna that provides an almost hemispherical coverage. The receiver can track either one or both the GPS code types, which will be described in Section 3.3. Most receivers have multiple channels, each tracking transmission from a single satellite.

A simplified block diagram of the multichannel receiver is shown in Figure 3.5. The received signals are usually filtered, to reduce out-of-band interference, preamplified and downconverted at intermediate frequency, and, increasingly often, sampled and digitized, exploiting oversampling techniques to reduce the analog-to-digital (A/D) converter complexity. Samples are then addressed to the digital signal processor (DSP) section, which contains N parallel channels to track simultaneously carriers and code from N satellites. Each channel includes code and carrier tracking loops to perform code and carrier phase measurements and demodulation of the navigation message data. Various satellite-to-user measurements can be performed, such as pseudorange (basic and delta) and integrated Doppler, that are forwarded—together with the demodulated navigation message —to the navigation/receiver processor. The latter is generally required to control and command the receiver, starting the operational sequence with signal acquisition and following with signal tracking and data collection [21–35].

An input/output (I/O) unit interfaces the GPS set to the user. In many applications the I/O device is a control display unit (CDU), which allows data entry, status, and navigation (positioning, velocity, and timing) information display and access to various navigation functions. As an example, Figures 3.6 and 3.7 show commercially available handheld and marine GPS terminals, respectively.

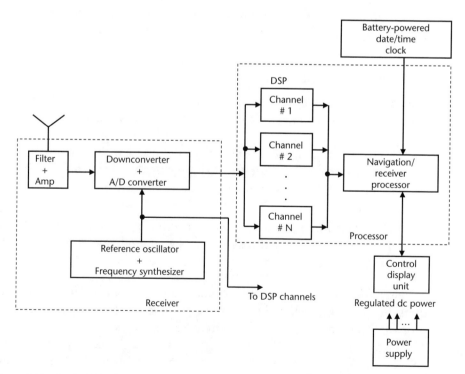

Figure 3.5   General block diagram of the GPS receiver.

GPS users belong to either military or civilian environments. GPS allows them to determine, without any direct charge, their position anywhere in the world and to exploit this information for several related applications.

Depending on the provided accuracy (see Chapter 2), GPS provides two levels of services: standard positioning service (SPS) and precise positioning service (PPS). Their accuracy specification standard will be defined in Section 3.5.

In particular, GPS can be exploited for *aircraft* navigation in commercial and general aviation applications; *land* mobile navigation for vehicles (cars, trucks, and buses); and *maritime* ship navigation. In addition, GPS can be used for time transfer between clocks, spacecraft orbit determination, attitude determination using multiple antennas, kinematics survey, and ionosphere measurement. Military

Figure 3.6 GPS commercial handheld terminal. (*From:* [36]. © 2004 Garmin Corporation. Reprinted with permission.)

Figure 3.7   GPS commercial marine terminal. (*From:* [36]. © 2004 Garmin Corporation. Reprinted with permission.)

applications include troop deployment and observation and target tracking for remote sensing applications and smart weapons. Civilian applications include vehicle tracking, *geographic information system* (GIS) data collection, emergency services, agriculture, photogrammetry, and recreation (e.g., hiking).

3.3 GPS Signals

A major requirement related to the signals of a satellite navigation system is the availability of simultaneous access to signals transmitted from a set of satellites with the minimum cross-interference [10, 12, 37, 38]. In GPS, this translates into the use of a code division multiple access (CDMA) scheme, where the signal coming from each satellite is modulated by its own pseudorandom noise (PRN) code, that has a low correlation with codes assigned to the signals transmitted from the other satellites. This technique allows the user to receive multiple signals on the same frequency band with a low mutual interference. The signal at the user receiver has a much wider spectrum than the one required from the navigation data and a very low power spectral density, below the thermal noise level. As a consequence, a noticeable resistance to interference and jamming is achieved together with a low detection capability from nonauthorized users.

In what follows, the original structure of the GPS signal will be described. It was designed in the early 1970s and, through the GPS modernization program (launches of block IIR, IIF, and III), it is being enriched to meet the current and envisaged user needs [39].

Each satellite is conceived for transmitting continuously the navigation message over two L-band carriers:

L1   at 1,575.42 MHz

L2   at 1,227.60 MHz

with:

$$L1 = 154f_0 \tag{3.1}$$

$$L2 = 120f_0 \tag{3.2}$$

where f_0 is a nominal reference frequency (also referred to as the fundamental frequency) generated onboard that appears to be 10.23 MHz to an observer on ground.

The L1 frequency is binary phase shift keying (BPSK) modulated by the coarse/acquisition code (C/A code) and by the precision code (P code) in quadrature, whereas the L2 frequency is only BPSK modulated by the P code.

The C/A codes are available to all civilian users and are the basis for the provision of the SPS. The C/A code is 1,023 chips long and is generated by properly combining the output taps of two 10-stage linear feedback shift register (LFSR) binary sequence generators. In particular, the final code is achieved by a module-2 addition (exclusive-or, XOR) between the output of one of the two registers and a delayed version of the output of the other one. The delay amount of the latter register identifies the different code sequences, determining, hence, the satellite identification. Furthermore, as the all-zero output among the 1,024 possible states of each register is discarded, a code composed of 1,023 bits is then achieved (*maximum length code*). Among the 1,023 possible codes, corresponding to the number of possible delays, only the codes that are mutually orthogonal, and hence noncorrelated (*gold codes*), are selected, in order to reduce the cross-interference among the signals.

The rate of the code bit (*chip* rate) is $R_{C/A} = 1.023$ Mchip/s and, having an $L_{C/A} = 1,023$ code length, the code duration is $L_{C/A}/R_{C/A} = 1$ ms. It is worth noting that $R_{C/A} = f_0/10$.

Contrary to the C/A codes, the access to the P codes can be denied to the SPS users, through activation by the control segment of the *antispoofing* (AS) mode. In fact, P codes are mainly conceived for military purposes and, although they were available to all users until 1994, they can since be denied by adding an unknown encryption code to them. The encrypted version of the P code is indicated as *Y code* and has the same chip rate as the P code. For this reason the P code is often indicated as *P(Y) code*.

P(Y) codes are accessible only to DoD-authorized users, such as U.S. militaries, NATO, and selected military task forces, that can be hence admitted to the PPS. The latter can be rendered available to civilian users only under DoD permission.

P codes adopt a very long sequence, with chip rate $R_P = 10.23$ Mchip/s. It is worth noting that $R_P = f_0$. The full P code stream has a repetition time of about 266 days (i.e., 38 weeks), which results in 2.35×10^{14} chip length. The 266-day-long code is divided into 38 segments that correspond to 38 weekly sections of the code. Each segment, hence, results in a 7-day-long code with, considering the chip rate R_P, a code length L_P of 6.1871×10^{12} chips. The segments are assigned to the GPS constellation satellites (one per satellite) and the remaining are assigned to other users. Each GPS satellite is identified by the number i that corresponds to the ith segment of the P code that is uniquely assigned to the satellite itself.

The *navigation signal* is composed of:

- Carrier (L1, L2);
- Code [C/A, P(Y)];
- Navigation data (at a bit rate $R_{ND} = 50$ bps).

The navigation data are the core of the GPS signal. These data bits are organized in 5 subframes, each containing 300 bits, and with a duration of 6 seconds. The navigation data include precise satellite ephemeris as a function of time, precise satellite clock parameters, atmospheric data, and an almanac. In particular, the ephemeris parameters are a precise fit to the transmitting satellite orbits and are valid only for a few hours; the almanac is a reduced precision subset of the ephemeris that is exploited to predict the approximate satellite position and provide an aid in satellite signal acquisition. Almanac includes orbital data, low-precision clock data, simple configuration, and health status for every satellite of GPS constellation.

In addition navigation data includes other information useful for the user signal acquisition, such as user messages, ionospheric model data, and UTC calculations.

Figure 3.8 shows the block diagram of the GPS satellite signal structure. Because both carriers and code chip rates are multiples or submultiples of the reference frequency f_0, the navigation signal components are all synchronized.

As a result of the described composition and features of the GPS navigation signals, the signal $x_{L1,k}(t)$ transmitted by the kth GPS satellite on the L1 carrier can be expressed in the form:

$$x_{L1,k}(t) = \sqrt{2P_{c/a}}\, C_{C/A,k}(t)D_k(t)\cos(\omega_1 t + \varphi) + \sqrt{2P_p}\, C_{P,k}(t)D_k(t)\sin(\omega_1 t + \varphi)$$

(3.3)

where $P_{c/a}$ and P_p are the C/A and P signal powers, respectively; $C_{C/A,k}(t)$ is the kth satellite unique gold code for the C/A coding; $C_{P,k}(t)$ is the kth satellite PRN sequence of a 1-week period for the P coding; and $D_k(t)$ is the binary sequence of the navigation data.

The signal $x_{L2,k}(t)$ transmitted by the kth GPS satellite on the L2 carrier can be, instead, expressed in the form:

$$x_{L2,k}(t) = \sqrt{2P_{L2}}\, C_{P,k}(t)D_k(t)\cos(\omega_2 t + \varphi)$$

(3.4)

where P_{L2} is the signal power and the other symbols have the same meaning as in (3.3). The GPS system uses one-way satellite ranging and trilateration to determine

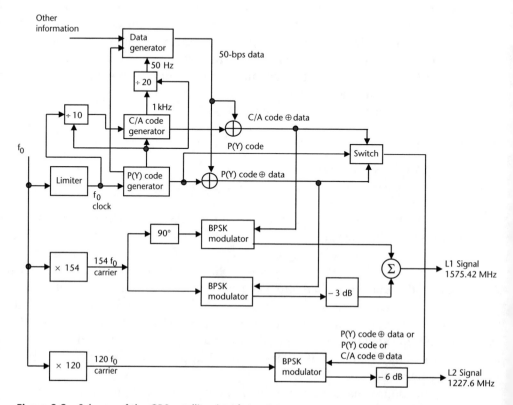

Figure 3.8 Scheme of the GPS satellite signal structure.

the position of a GPS receiver on the Earth's surface; the GPS reference system is WGS-84 (see Chapter 2).

The range measurement from the transmitted GPS satellite signals can be achieved through a code tracking technique within the receiver. The received signal, after translation to baseband, is correlated with a locally generated code sequence, related to a selected satellite of the constellation. The receiver analyzes the correlation peak. Thanks to the properties of the code sequences (C/A code), the signals coming from satellites other than the one under consideration can be discarded.

The time shift that has to be imposed to the locally generated code in order to be synchronized with the received signal, hence achieving the correlation peak, provides the signal propagation time. The latter, if multiplied by the light speed, gives the measured range (pseudorange), which differs from the actual one because possible error sources (e.g., troposphere and ionosphere propagation, clocks) have not been taken into account in the process. In addition to the pseudorange measurements, satellite coordinates are required to determine the user position. This information is contained in the navigation message, which is decoded from the satellite signal by exploiting the locally generated code.

The P(Y) code acquisition—for an authorized user—takes place through the C/A code. In fact, the receiver first acquires the C/A code and then the P(Y) code, through time information included in the data message. In fact, the direct acquisition of the P(Y) code is difficult, due to the extremely high code length, as this would require a very high precision clock and thousands of parallel correlators. An alternative technique for range and Doppler shift measurements is the carrier tracking: the receiver generates locally the carrier frequency and exploits this signal for the range measurement. The latter can be achieved by comparing the phase difference between the local carrier—properly synchronized with the system time—and the satellite transmitted carrier, which has undergone the propagation in the radio channel. Although those measures are very precise, their accuracy is limited by the ambiguity on the received carrier cycle (cycle ambiguity), since the receiver cannot directly determine the exact number of whole cycles in the pseudorange. The latter problem can be overcome by using carrier-phase differential GPS, which provides carrier-phase measurements to nearby user receivers through ground reference stations, located in known points. In particular, ground stations use differential techniques that, properly combining the GPS signals, resolve the cycle ambiguity problem, improving, hence, the GPS error budget. Considering the current availability of two GPS signals, L1 and L2, two main techniques have been developed: *geometry-free* and *geometry-dependent*. The first one determines the whole-cycle ambiguities of the carrier-phase measurements by using smoothed code measurements. The second technique gives the "best" solution, according to some criteria (typically a minimum sum square of the residuals), of a search process that combines whole-cycle ambiguities [40]. Figures 3.9 and 3.10 show the block diagram of the GPS receiver code and carrier tracking loops, respectively.

The GPS modernization, which is described in detail in Chapter 6, implies the introduction of a civilian signal also on L2 frequency (L2C) and two novel Mcodes on L1 and L2 carriers. The availability of two civilian codes allows a GPS isolated user to improve performance, since, properly combining the two signals, they

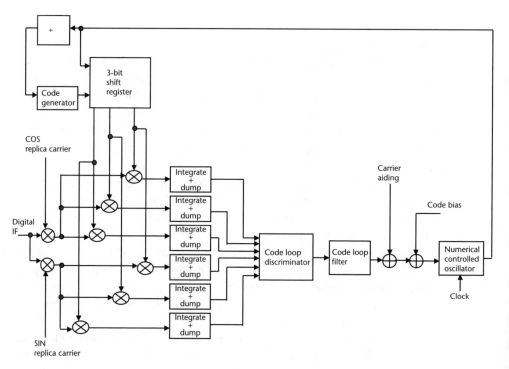

Figure 3.9   Basic GPS receiver code tracking loop.

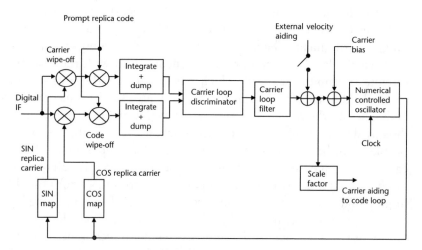

Figure 3.10   Basic GPS receiver carrier tracking loop.

partially correct some propagation impairments of the GPS signal due to ionosphere (see Section 3.4). M codes use spectrum portions already assigned to the L1 and L2 frequencies, although remaining spectrally distinguishable from the civilian codes at the same frequencies. This would meet the request, issued from the military environment, to develop countermeasures against a hostile use of GPS, allowing the United States and its partners to maintain military control over the system without affecting civilian user performance. These new codes benefit from new cryptography algorithms and from the local possibility of bringing the transmission power up to a 20-dB higher level than in current P(Y)-coded signals [39, 41].

This extension of the civilian code is, however, still insufficient for an extensive use of GPS in civil aviation. As a consequence, a new civil frequency (L5 = 1,176.45 MHz) will be added to the extended structure. A high power level and large broadcast bandwidth (20 MHz at least) will be assigned to L5 together with a high chip rate (f_0) that will guarantee higher accuracy under noise and multipath impairments. The broadcast navigation message on the novel signal will have a fully different and more effective structure. The code will be longer than the current C/A version, hence improving robustness to interference. In addition, referring to the cycle ambiguity problem already highlighted, the availability of a third GPS carrier can be exploited in the carrier-phase differential GPS (e.g., three-carrier phase ambiguity resolution (TCAR) method [42]).

3.4   Signal Propagation Effects

The atmosphere affects the GPS signal propagation, resulting in a delay that cannot be ignored in the GPS error budget (see Section 3.5). The impact of the atmospheric effects on the GPS satellite transmissions depends on the elevation angle and the user location environment [10, 12]. The L-band frequency range, where GPS satellites transmit, is high enough to keep the ionosphere delay effects quite small and low enough to be compliant with the use of small omnidirectional antennas and to have no significant loss due to rain. Nonetheless, the atmosphere causes nonnegligible effects that include ionospheric group delay and scintillation; group delay caused by wet and dry atmosphere (troposphere and stratosphere); atmosphere attenuation in the troposphere and stratosphere; and multipath from reflective surfaces and scattering [43–49].

For GPS application, the atmospheric regions of interest for error budget are mainly two: troposphere and ionosphere. Troposphere ranges from the surface of the Earth from 12 to 20 km in altitude, whereas ionosphere ranges from 75 to 500 km.

The ionosphere is characterized by a large concentration of free electrons that cause a variation of the channel refraction index and, consequently, of the wave propagation velocity in the medium. In particular, the ionosphere effects on the GPS signals mainly result in either a combination of group delay and carrier phase advance, which varies with the paths and density encountered by the signal in crossing the ionosphere, or scintillation, which can bring the signal amplitude and phase to fluctuate rapidly at certain latitudes and even cause loss of lock. Other effects related to the ionosphere, such as Faraday rotation and ray bending that changes the arrival angle, are not significant for the GPS frequency band. In the GPS frequency ranges, the ionospheric zenith path delay might vary from 2 to 50 ns, in a diurnal fashion that is quite variable and unpredictable from day to day. At lower elevation angles, but high enough for the ionosphere to be penetrated by the signal, the signal path crosses a larger extent of the ionosphere. The delay might increase up to three times the figure of the zenithal penetration. This implies that the zenithal delay of 50 ns translates into a 150 ns delay corresponding to a range error of 45m.

The latter figure is clearly not consistent with the required accuracy of the GPS system.

Compared with the ionospheric effect, the tropospheric effect is about an order of magnitude less. This region has a refraction index that varies with the altitude. Because the tropospheric refraction index is greater than one, the crossing of this region causes a group delay in the GPS signals. Different from the ionosphere, this delay does not depend on the signal carrier. The troposphere induces attenuation effects that are generally below 0.5 dB and delay effects in the order of 2–25m. These effects vary with elevation angle (as lower angles bring to a longer path length through the troposphere), and the atmospheric gas density profile. It should be underlined also that about one-fourth of the delay effect is caused by atmospheric gases above the troposphere, specifically in the tropopause and stratosphere.

At GPS frequencies, oxygen is the main source of attenuation (0.035-dB zenithal attenuation at 1.5 GHz), whereas the effects of water vapor, rain, and nitrogen attenuation are negligible. Moving from the zenithal penetration, the impact of an oblique penetration of troposphere at low elevation angles is much larger than that of the ionosphere (a factor greater than 3), because the troposphere extends down to the Earth's surface. As a consequence, the use of GPS satellites should be avoided below approximately 5° elevation: this is advisable not only to avoid higher attenuation but also because of the larger uncertainties in the tropospheric delay and the greater scintillation effects. In addition, at low elevation angles other effects, such as reflections, refraction, and receiving antenna gain roll-off, might be amplified.

Tropospheric scintillation, caused by irregularities and turbulence in the atmospheric refractive index, varies with time and other parameters, such as frequency, elevation angle, and weather conditions (in particular, dense clouds). At GPS frequencies, these effects are quite small, except for a reduced fraction of time and at low elevation angles.

In addition, it is important to highlight the multipath effect, which can be critical, particularly for aircraft navigation. The multipath propagation phenomenon results in arriving at the receiving antenna of radio signals by two or more paths, due to multiple reflections over the satellite-receiver path. The effect of several paths causes the presence of multiple phased and attenuated replies of the main signal (referring to the signal with the line of sight path) at the receiver side. Therefore, these signals cause constructive and destructive interference in the receiving antenna, affecting, hence, the received signal quality. Nonetheless, the GPS receiver can effectively reject multipath under certain conditions. Within the available bandwidth and the constraints for the receiver complexity, the GPS signal is designed to resist the interference from multipath signals with mutual delay differences exceeding 1 μs. In addition, a further gain can be obtained by using specific signal processing software or devices in the receiving antennas (e.g., ground planes). In particular, the use of ground planes can shield the antenna element from the undesired signals before they reach the receiver, hence improving the received signal level [47–50]. Therefore, error compensation is required based on either modeling, measurement operations, or differential techniques.

The GPS receivers have difficult or very diminished reception in environments such as indoors, underground, under heavy tree canopy, urban and natural canyons, as well as around strong radio transmissions or power radio transmitter antennas.

All of these effects, in particular referring to the ionospheric ones, affect the GPS performance, resulting in a delay contribution in the GPS error budget, as highlighted in the following section.

3.5   GPS Performance

Although conceived for military purposes, satellite navigation is already making a very significant impact in civilian applications. The GPS system is influencing many aspects of everyday life, including navigation, precision landing, emergency services for mobile users, precision agriculture, and support in excursion and sports. The initial GPS specifications for both SPS and PPS users are shown in Table 3.1 [51].

An increase in the system accuracy together with an improvement in system availability and integrity can be achieved by means of additional signals. This will be achieved with the next satellite block (see Chapter 6), that, referring to accuracy, will allow an average accuracy of the GPS system around 10m. To this respect, Table 3.2 shows the last SPS accuracy specification standard defined by DoD on October 4, 2001 [52].

Until 2000, full GPS accuracy was made selectively available by an intentional degradation of the SPS user navigation solution. SA, operated by the DoD [53, 54], was implemented in 1990 and consisted of an intentional manipulation of the broadcast ephemeris data and the *dithering* of the satellite clock. This translated into a time-varying disturbance in the pseudorange and carrier-phase measurement and in the navigation data. Clock errors introduced random oscillation in a 4-to- 12-minute period and variations in the pseudorange error up to 70 m. SA was deactivated in May 2000, hence eliminating the major source on positioning error.

GPS system performance, in both the basic and modernized versions, is mainly judged by users in terms of *accuracy*, which implies the degree of conformance between measured and true positioning, velocity, and timing information. This accuracy depends on a complicated interaction among different factors [55–57]. In

Table 3.1   Initial GPS Performance (95% Probability)

User	Horizontal Accuracy (m)	Vertical Accuracy (m)	Speed (m/s)	Time (ns)
SPS	100	300	≤ 2	340
PPS	22	27.7	0.2	200

Table 3.2   SPS Positioning and Timing Accuracy Standard (95% Probability)

	Horizontal Error (m)	Vertical Error (m)	Time Transfer Error (ns)
Global average positioning domain accuracy	≤ 13	≤ 22	≤ 40
Worst site positioning domain accuracy	≤ 36	≤ 77	≤ 40

general, GPS accuracy depends on the quality of the pseudorange measurements as well as on the satellite ephemeris data. In addition, other parameters, such as the accuracy of the satellite clock offsets relative to the GPS system time and of the satellite downlink propagation error estimation, are of key importance for determining the user accuracy performance. Noticeable error sources are included in the space, control, and user segment of the GPS system. It is generally assumed that error sources can be allocated to individual satellite pseudoranges and considered as effectively resulting in an equivalent error in the pseudorange figures [known as user equivalent range error (UERE)].

For a given satellite, the UERE is considered to be the statistical sum of the contributions from each of the error sources associated with the satellite. The contributions are generally considered independent, and the overall UERE for a satellite is approximately a zero-mean Gaussian random variable with variance given by the sum of each component variance. The UERE is usually considered independent and identically distributed (i.i.d.) from satellite to satellite.

The error E_{GPS} in the GPS solution can be estimated by:

$$E_{GPS} = GF \times E_{PSR} \tag{3.5}$$

where GF is a geometry factor that expresses the composite effect of the relative satellite-user geometry on the GPS solution error and it is often indicated as dilution of precision (DOP) associated with the satellite-user geometry; E_{PSR} is the pseudo-range error factor that, under proper assumptions, becomes the satellite UERE.

Several geometry factors can be defined and used in the estimation of the various components of the GPS navigation solution. The most general parameter is termed the *geometric* DOP (GDOP). Other DOP parameters can be usefully exploited to characterize the accuracy: *position* DOP (PDOP), *horizontal* DOP (HDOP), *vertical* DOP (VDOP), and *time* DOP (TDOP). These DOP parameters are defined in terms of the satellite UERE and elements of the covariance matrix of the position/time solution.

The effects of satellite and receiver clock offsets, ephemeris prediction errors, and various error sources degrade the satellite-to-user range measurement, bringing to the previously mentioned E_{PSR} pseudorange error. The satellite signal is also delayed when propagating through the atmosphere (ionosphere, troposphere). In addition, other contributions to the overall error are given by reflections (i.e., multipath effect), receiver noise and resolution effect, and receiver hardware offsets. All these contributions result in a total time offset that can be expressed as the sum of all delay/offset components.

The error budget for the GPS horizontal positioning performance is given in Table 3.3, where the impact of the SA degradation is outlined. After the SA removal, the major contribution to the overall error budget is given by the delay introduced by the ionosphere. The availability of the code on two carriers (i.e., the military users of the P code for basic GPS and all users for the modernized system) allows correction of the ionosphere delay through a proper combination of the L1 and L2 signals [43, 58–60].

Some manufacturers have developed innovative techniques that allow civilian users of the basic GPS to partially use the P(Y) code and, hence, evaluate the

Table 3.3   Horizontal GPS Errors for the SPS (C/A Code) at 95%

Error Source	Error (m) with SA	Error (m) Without SA
SA	24	0
Atmospheric delay		
Troposphere	0.2	0.2
Ionosphere	7.0	7.0
Ephemeris and clock	2.3	2.3
Receiver noise	0.6	0.6
Multipath	1.5	1.5
Total UERE	25.0	7.5
HDOP (typical)	1.5	1.5
Overall horizontal accuracy	75.0	22.5

ionosphere effects. Nonetheless, those techniques are not effective enough in the presence of low signal-to-noise ratio, as in the case of mobile users and of atmospheric scintillation effects.

The use of two signals, instead, results in a noticeable improvement in the GPS accuracy performance, as shown in Table 3.4 [51]: the availability of the C/A code also on the L2 frequency, as envisaged in modernized GPS, will be key to improving system performance for all civilian users and related applications.

The availability of the L5 frequency, which is envisaged in the advanced GPS modernization plan, would further improve accuracy, allowing, for instance, the GPS to support aviation in the whole en route flight phase [61]. After implementation of L5, the major contribution to the GPS error budget will be given by clock and ephemeris inaccuracy. An innovative technique, the *accuracy improvement initiative* (initiated and approved in 1996 by DoD), allows users to reduce those errors by about 50% by keeping into account the data coming from additional monitoring stations. The resulting accuracy budget is shown in Table 3.5 [51].

The improvement of GPS accuracy is strongly related to the number of reference stations on the ground, hence to the average distance between the user and the

Table 3.4   Horizontal GPS Errors for SPS with Two Signals (i.e., C/A Code Also on L2, at 95%)

Error Source	Error (m)
SA	0
Atmospheric delay	
Troposphere	0.2
Ionosphere	0.1
Ephemeris and clock	2.3
Receiver noise	0.6
Multipath	1.5
Total UERE	2.8
HDOP (typical)	1.5
Overall horizontal accuracy	8.5

Table 3.5 Horizontal GPS Errors for the SPS
with Additional Stations at 95%

Error Source	Error (m)
SA	0
Atmospheric delay	
Troposphere	0.2
Ionosphere	0.1
Ephemeris and clock	1.25
Receiver noise	0.6
Multipath	1.5
Total UERE	2.0
HDOP (typical)	1.5
Overall horizontal accuracy	6.0

closest reference station. In fact, the presence of a station allows one to delete the error component that is correlated between a user and station. As discussed in Chapter 4, the concept of system augmentation shows that performance can also be improved through an increase of the density of the on-ground reference stations.

References

[1] El-Rabbany, A., *Introduction to GPS: The Global Positioning System*, Norwood, MA: Artech House, 2002.

[2] "GPS: The Global Positioning System," Special Issue, *IEEE Proceedings*, Vol. 87, No. 1, January 1999.

[3] Lasiter, E. M., and B.W. Parkinson, "The Operational Status of NAVSTAR GPS," *Journal of Navigation*, Vol. 30, No. 1, 1977.

[4] Easton, R. L., "The Navigation Technology Program: Global Positioning System," *ION*, Vol. I, Washington, D.C., 1980, pp. 15–20.

[5] Baker, P. J., "GPS in the Year 2000 and Beyond," *Journal of Navigation*, Vol. 40, No. 2, 1987.

[6] Parkinson, B. W., "Overview, Global Positioning System," *ION*, Vol. I, Washington D.C., 1980, p. 1.

[7] Burgess, A., "GPS Program Status," *Proc. of Nav 89: Satellite Navigation*, London, England: Royal Institute of Navigation, 1989.

[8] Shirer, H., "GPS and the U.S. Federal Radionavigation Plan," *GPS World*, Vol. 2, No. 2, 1991.

[9] Baker, P. J., "GPS Policy." *Proc. of the Fourth International Symposium on Satellite Positioning*, University of Texas at Austin, 1986.

[10] Kaplan, E. D., *Understanding GPS: Principles and Applications*, Norwood, MA: Artech House, 1996.

[11] Hoffmann-Wellenhof, B., H. Lichtenegger, and J. Collins, *Global Positioning System: Theory and Practice*, 3rd ed., New York: Springer-Verlag, 1994.

[12] Parkinson, B. W., and J. J. Spilker Jr., (eds.), "Global Positioning System: Theory and Applications," *Progress in Astronautics and Aeronautics*, American Institute of Aeronautics and Astronautics, Vols. 163 and 164, 1996.

[13] Langley, R. B., "The Orbits of GPS Satellites," *GPS World*, Vol. 2, No. 3, March 1991, pp. 50–53.

[14] Leick, A., *GPS Satellite Surveying*, 2nd ed., New York: Wiley, 1995.

[15] Maine, K. P., P. Anderson, and J. Lauger, "Cross-Links for the Next Generation GPS," *Proc. IEEE Aerospace Conference*, Big Sky, MT, March 2003, Paper No. 4.1302.

[16] "GPS Status," U.S. Coast Guard Navigation Center, September 2001, http://www.navcen.uscg.gov/gps/.

[17] Langley, R. B., "The GPS Receiver: An Introduction," *GPS World*, Vol. 2, No. 1, January 1991, pp. 50–53.

[18] Langley, R. B., "Smaller and Smaller. The Evolution of the GPS Receiver," *GPS World*, Vol. 11, No. 4, April 2000, pp. 54–58.

[19] Bao-Yen Tsui, J., *Fundamentals of Global Positioning System Receivers: A Software Approach*, New York: Wiley-Interscience, 2000.

[20] Abel, J. S., and J. W. Chaffee, "Existence and Uniqueness of GPS Solutions," *IEEE Trans. on Aerospace and Electronic Systems*, Vol. 27, No. 6, November 1991, pp. 952–956.

[21] Fang, B. T., "Comments on 'Existence and Uniqueness of GPS Solutions' by J. S. Abel and J. W. Chaffee," *IEEE Trans. on Aerospace and Electronic Systems*, Vol. 28, No. 4, October 1992, p. 1163.

[22] Leva, J. L., "An Alternative Closed-Form Solution to the GPS Pseudorange Equations," *IEEE Trans. on Aerospace and Electronic Systems*, Vol. 32, No. 4, October 1996, pp. 1430–1439.

[23] Chaffee, J. W., and J. S. Abel, "The GPS Filtering Problem," *Proc. IEEE PLANS*, (Record. "500 Years After Columbus—Navigation Challenges of Tomorrow"), Las Vegas, NV, March 1992, pp. 12–20.

[24] Mao, X., M. Wada, and H. Hashimoto, "Investigation on Nonlinear Filtering Algorithms for GPS," *Proc. IEEE Intelligent Vehicle Symposium*, Paris, France, Vol. 1, June 2002, pp. 64–70.

[25] Mao, X., M. Wada, and H. Hashimoto, "Nonlinear Filtering Algorithms for GPS Using Pseudorange and Doppler Shift Measurements," *Proc. IEEE 5th Intelligent Transportation Systems Conf.*, Singapore, 2002, pp. 914–919.

[26] Wu, S. C., and W. G. Melbourne, "An Optimal GPS Data Processing Technique for Precise Positioning," *IEEE Trans. on Geoscience and Remote Sensing*, Vol. 31, No. 1, January 1993, pp. 146–152.

[27] Nardi, S., and M. Pachter, "GPS Estimation Algorithm Using Stochastic Modeling," *Proc. IEEE 37th Decision and Control Conf.*, Vol. 4, Tampa, FL, December 1998.

[28] Chaffee, J., J. Abel, and B. K. McQuiston, "GPS Positioning, Filtering, and Integration," *Proc. IEEE NAECON 1993*, Dayton, OH, May 1993, Vol. 1, pp. 327–332.

[29] Garrison, J. L., and L. Bertuccelli, "GPS Code Tracking in High Altitude Orbiting Receivers," *Proc. IEEE PLANS*, Palm Springs, CA, April 2002, pp. 164–171.

[30] Weihua, Z., and J. Tranquilla, "Modeling and Analysis for the GPS Pseudorange Observable," *IEEE Trans. on Aerospace and Electronic Systems*, Vol. 31, No. 2, April 1995, pp. 739–751.

[31] Chaffee, J. W., "Observability, Ensemble Averaging and GPS Time," *IEEE Trans. on Aerospace and Electronic Systems*, Vol. 28, No. 1, January 1992, pp. 224–240.

[32] Xu, G., *GPS: Theory, Algorithms, and Applications*, New York: Springer-Verlag, 2003.

[33] Akopian, D., and J. Syrjarinne, "A Network Aided Iterated LS Method for GPS Positioning and Time Recovery Without Navigation Message Decoding," *Proc. IEEE PLANS*, Palm Springs, CA, April 2002, pp. 77–84.

[34] Kokkoninen, M., and S. Pietila, "A New Bit Synchronization Method for a GPS Receiver," *Proc. IEEE PLANS*, Palm Springs, CA, April 2002, pp. 85–90.

[35] Pathak, M. S., "Recursive Method for Optimum GPS Selection," *IEEE Trans. on Aerospace and Electronic Systems*, Vol. 37, No. 2, April 2001, pp. 751–754.

[36] http://www.garmin.com.

[37] Langley, R. B., "Why Is the GPS Signal So Complex?" *GPS World*, Vol. 1, No. 3, May/June 1990, pp. 56–59.

[38] Spilker Jr., J. J., "GPS Signal Structure and Performance Characteristics," *Navigation*, Vol. 25, No. 2, 1978.

[39] Shaw, M., K. Sandhoo, and D. Turner, "Modernization of the Global Positioning System," *GPS World*, Vol. 11, No. 9, September 2000, pp. 36–44.

[40] Abidin, H. Z., "Multi-Monitor Station On-the-Fly Ambiguity Resolution: The Impacts of Satellite Geometry and Monitor Station Geometry," *Proc. of IEEE PLANS '92*, Monterey, CA, March 1999, pp. 412–418.

[41] Barker, B. C., et al., "Details of the GPS M Code Signal," *Proceedings of ION 2000 National Technical Meeting*, Institute of Navigation, January 2000.

[42] Volath, U., et al., "Analysis of Three-Carrier Phase Ambiguity Resolution (TCAR) Technique for Precise Relative Positioning in GNSS-2," *Proc. of ION GPS-98*, Nashville, TN, September 1998.

[43] Klobuchar, J. A., "Ionospheric Effects on GPS," *GPS World*, Vol. 2, No. 4, April 1991, pp. 48–51.

[44] Langley, R. B., "GPS, the Ionosphere and the Solar Maximum," *GPS World*, Vol. 11, No. 7, July 2000, pp. 44–49.

[45] Brunner, F. K., and W. M. Welsch, "Effect of the Troposphere on GPS Measurements," *GPS World*, Vol. 4, No. 1, January 1993, pp. 42–51.

[46] Hay, C., and J. Wong, "Enhancing GPS: Tropospheric Delay Prediction at the Master Control Station," *GPS World*, Vol. 11, No. 1, January 2000, pp. 56–62.

[47] Weill, L. R., "Conquering Multipath: The GPS Accuracy Battle," *GPS World*, Vol. 8, No. 4, April 1997, pp. 59–66.

[48] Kelly, J. M., and M. S. Braasch, "Validation of Theoretical GPS Multipath Bias Characteristics," *Proc. of IEEE Aerospace Conference*, Big Sky, MT, Paper No. 4.1103, March 2001.

[49] Ray, J. K., M. E. Cannon, and P. Fenton, "GPS Code and Carrier Multipath Mitigation Using a Multi-Antenna System," *IEEE Trans. on Aerospace and Electronic Systems*, Vol. 37, No. 1, January 2001, pp. 183–195.

[50] McKinzie, W. E., et al., "Mitigation of Multipath Through the Use of an Artificial Magnetic Conductor for Precision GPS Surveying Antennas," *IEEE Antennas and Propagation Society International Symposium*, Vol. 4, San Antonio, TX, June 2002, pp. 640–643.

[51] Galati, G., *Detection and Navigation Systems/Sistemi di Rilevamento e Navigazione*, Italy: Texmat, 2002.

[52] "Global Positioning System Standard Positioning Service Performance Standard," U.S. Department of Defense, October 2001.

[53] Georgiadou, Y., and K. D. Doucet, "The Issue of Selective Availability," *GPS World*, Vol. 1, No. 5, September/October 1990, pp. 53–56.

[54] Conley, R., "Life After Selective Availability," *U.S. Institute of Navigation Newsletter*, Vol. 10, No. 1, Spring 2000, pp. 3–4.

[55] Kleusberg, A., and R. B. Langley, "The Limitations of GPS," *GPS World*, Vol. 1, No. 2, March/April 1990, pp. 50–52.

[56] Langley, R. B., "Time, Clocks and GPS," *GPS World*, Vol. 2, No. 10, November/December 1991, pp. 38–42.

[57] Dong-Ho, S., and T. K. Sung, "Comparison of Error Characteristics Between TOA and TDOA Positioning," *IEEE Trans. on Aerospace and Electronic Systems*, Vol. 38, No. 1, January 2002, pp. 307–311.

[58] Afraimovich, E. L., V. V. Chernukhov, and V. V. Dernyanov, "Updating the Ionospheric Delay Model Using GPS Data," Application of the Conversion Research Results for International Cooperation, *Third International Symposium, SIBCONVERS '99*, Vol. 2, Tomsk, Russia, May 1999, pp. 385–387.

[59]   Batchelor, A., P. Fleming, and G. Morgan-Owen, "Ionospheric Delay Estimation in the European Global Navigation Overlay Service," *IEEE Colloquium on Remote Sensing of the Propagation Environment*, Digest No. 1996/221, November 1996, pp. 3/1–3/6.

[60]   Kovach, K., "New User Equivalent Range Error (UERE) Budget for the Modernized Navstar Global Positioning System (GPS)," *Proc. of The Institute of Navigation National Technical Meeting*, Anaheim, CA, January 2000.

[61]   Hatch, R., et al., "Civilian GPS: The Benefits of Three Frequencies," *GPS Solutions*, Vol. 3, No. 4, 2000, pp. 1–9.

CHAPTER 4

Augmentation Systems

4.1 Introduction

Chapter 3 delineated the many interesting and effective features of the best testimonial of the navigation world: GPS. Nonetheless, stand-alone GPS cannot be used for applications, like civil aviation, where security is a critical task. This is due to both the nonadequate continuity of the location data, for instance during the precision approach and instrumental landing phases, and the lack of real-time information to the user about the provided quality of service (*integrity*). The same applies to the GLONASS system.

An interesting report of a few years ago [1] investigated the ability of GPS to meet the required navigation performance in terms of accuracy, integrity, continuity, and availability, in the frame of the transition, planned by the Federal Aviation Administration (FAA), from a ground-based navigation and landing system to a satellite-based approach exploiting GPS signals. A variety of risks for GPS performance were evaluated, related to multipath, ionosphere and troposphere, satellite ephemeris, unscheduled satellite failures, satellite unavailability due to scheduled maintenance, repair, repositioning or testing, and loss of ground support functions (i.e., health of the operational and master control stations and their associated communications functions) [2–7]. Signal emissions from other normal and expected transmissions were also evaluated in terms of potential interference with GPS signal reception, together with intentional interference sources. Ionosphere and interference risks were found to be the most significant ones. Among unintentional interference sources, commercial VHF radio, over-the-horizon military radars, and broadcast television have been mainly considered, together with sources of intentional interference caused by individuals or small groups (*hackers*) that exploit a technological weakness or by hostile organizations or governments that view the reliance of civil aviation on GPS as an opportunity for terrorist actions.

The anomalies caused by either satellite or the MCS may result in unpredictable range errors above the operational tolerances, which are more significant than the degraded accuracy due to poor satellite geometry. Integrity anomalies are rare: GPS should undergo no more than an average of three service failures per year. This failure rate is conservative and based on a historical estimation of Block II satellite and control segment failure characteristics [8]. However, they can be severe and catastrophic, particularly in air navigation [9, 10].

Among the main causes of integrity anomalies are satellite clock and ephemeris, such as in GPS Block I satellites, due to frequency standard problems (random runoff, large jump, and so on); the MCS clock jumps in the presence of large variations of the beam current or temperature of the frequency standard. Ephemeris anomalies occurred in Block I satellites in the process of the solar panel attempting to realign itself to track the sun after an eclipse [9]. Large ranging errors resulted from satellite attitude adjustment by means of firing the thrusters. Succeeding satellite versions removed this capability and introduced radiation hardening against the space environment. The lack of radiation hardening in Block I satellites resulted in ranging errors of thousand of meters in a few minutes. Errors of the same order of magnitude have been registered as a consequence of the loss of P code tracking. MCS anomalies are related to hardware, software, or human errors that result in errors of several thousands meters. Furthermore, the coverage of the GPS ground monitoring network is not spatially and temporally continuous, making it impossible to provide a system level real-time integrity monitoring for civilian applications. Hardware redundancy, software robustness, and training provision to prevent from human errors have been introduced to minimize possible integrity anomalies at the MCS. However, the resulting response times (time-to-alert) is still not sufficient for safety critical applications, such as aviation [9, 10], whose operations require from a GNSS the time-to-alert values reported in Table 4.1 [3, 11].

User-independent techniques have been developed to detect satellite anomalies, when occurring, and to overcome the nonsufficient continuity of the positioning data in safety critical applications. In addition, the GPS architecture has been interfaced and extended to create augmented systems able to overcome the performance limitations of stand-alone GPS and make satellite navigation systems the key support to safety critical applications, in particular aviation.

The following describes the evolution of stand-alone GPS, highlighting the achievements and many ongoing developments of this fascinating and dynamic mosaic.

4.2 Integrity Monitoring

As previously highlighted, GPS failures occur at each stage of the whole GPS operation. As a consequence, it is necessary to distinguish among failures that are induced at either system or user or operational environment level.

Table 4.1 GNSS Aviation Operational Time-to-Alert Requirements

Operation	Time to Alert
Oceanic	2 min
En route	1 min
Terminal	30 sec
NPA (nonprecision approach)	10 sec
APVI (approach with vertical guidance I)	10 sec
APVII (approach with vertical guidance II)	6 sec
CAT. I	6 sec
CAT. II	1 sec
CAT. III	1 sec

System level failures concern space and control segments as well as the interface between the two. The failures can be related to erroneous clock behavior, incorrect modeling and malfunction of the MCS, and anomalous performance of satellite payload, space vehicle and radio frequency (RF) sections.

User level failures are related both to user receiver equipment (hardware and software) and user operation (for example, lack of adequate training or overreliance on a single navigation system) [12].

Finally, operational environment-induced failures can be classified as intended, unintended, and signal propagation ones.

In the last decade, the FAA has devoted significant resources to the development of integrity provision techniques for airborne use, with the aim of making GPS the primary navigation system in aviation. This strategic approach and related achievements has resulted in the conception and development of ongoing and future navigation systems worldwide. This will make it possible to achieve a global air traffic management strategy as desired by the ICAO, based on satellite navigation and communication systems.

Receiver autonomous integrity monitoring (RAIM) is one of the approaches developed to provide integrity monitoring [9, 13]. The RAIM algorithm, which autonomously (i.e., at the receiver) exploits a determined solution of the navigation equations to check consistency, is embedded in the user terminal. Satellite anomalies are detected by exploiting at least five satellites in the navigation solution.

The approach is similar to that of "error detecting codes": the code is able to detect errors, warning the system properly, but not to correct them. A satellite anomaly is detected and a warning (alarm or alert) given to the user to trigger a countermeasure. For instance, a pilot could switch to an alternative navigation system to fly an aircraft.

The step forward for a transmission code is to become an "error correcting code" (i.e., to be able to detect a certain number of errors and correct part of them). Similarly, the step forward for an integrity monitoring technique is to detect the anomaly but also to remove the faulty satellite from the navigation solution, hence, identifying which satellite is not working properly (fault detection and *isolation*, or FDI, approach) or identifying another set of satellites that does not include the anomalous satellite (fault detection and *exclusion*, or FDE, approach).

In case GPS is the sole means of navigation, N satellites in visibility, with N greater or equal to six, are needed by the RAIM algorithm. If the FDI method is adopted, the detection of an integrity problem causes to form N solutions of $N-1$ satellites each, hence isolating the faulty satellite and removing it from the navigation solution. On the contrary, if the FDE method is chosen, the algorithm works with a set of six satellites, even if N is greater than six: if an integrity fault occurs, the algorithm searches for another set of six satellites that passes the consistency test, hence excluding the anomalous satellite from the navigation solution, without identifying it [9, 13].

In order to decide whether RAIM is available to the user receiver as an integrity monitoring technique, a lot of analyses were performed [3, 14–17]. They demonstrated that RAIM availability is not sufficient to meet the requirements for NPA and precision approaches (PA, from APV I down to CAT. III) flight phases, with respect to both horizontal alert limits and vertical alert limits (which are even more stringent), with the exception of some regions (for example, those in proximity of the equatorial line, but only for NPA and APV I flight phases in the case of horizontal requirements).

The purpose of illustrating the RAIM operation leads the authors to consider the horizontal performance requirements, which are also less stringent than the vertical ones and are satisfied by the RAIM method in some flight phases, such as oceanic and en route.

The RAIM basic algorithm has the goal of protecting the user against excessive horizontal position error, hence detecting if the horizontal error becomes higher than a certain threshold within a specified level of confidence. The position error cannot be observed directly, therefore another parameter has to be derived from what can be observed. That parameter is used as a mathematical indicator in a statistical method having the purpose of detecting a satellite failure. The basic algorithms move from the standard deviation of the measurement noise, the measurement geometry, and the maximum allowable probability of false alarm P_{fa} and missed detection P_{md}; as an output, the algorithm provides a protection parameter (horizontal protection limit, or HPL), that defines the smallest horizontal position error that can be detected for the specified P_{fa} and P_{md}.

The application of the algorithm to aviation implies the comparison of the HPL parameter with the HAL_i ICAO parameter, which varies according to the ongoing ith phase of flight. If:

$$HPL < HAL_i \tag{4.1}$$

RAIM is said to be available for the ith phase of flight. If condition (4.1) is not verified and GPS is used as secondary aid, the pilot switches to the primary navigation system.

The HPL parameter can be expressed as the product of three parameters:

$$HPL = S_{max} \sigma_{UERE} S_x \tag{4.2}$$

where σ_{UERE} is the standard deviation of the satellite pseudorange error (see Chapter 3), S_x is the density function parameter of the random variable that is used in modeling the test statistics and, indicating with N the number of measurements from visible satellites:

$$S_{max} = \max(S_j) \tag{4.3}$$
$$\text{for } j = 1, 2, ..., N$$

where S_j is the jth satellite slope of the characteristic slope line formed by the estimated horizontal position error versus the test statistics. Slopes are a function of the so-called "linear connection matrix" of the GPS measurement equation, which includes geometry and clock state information and varies slowly with time as satellites move along their orbits.

The fault detection capability of RAIM requires at least *five* visible satellites, whereas at least *six* satellites are required for fault detection and isolation or exclusion: therefore, a number of satellites higher than the minimum needed for the navigation function (i.e., *four*) is demanded by RAIM. As a result, RAIM has a lower availability than the navigation function.

In aviation, availability can be increased by including additional measurements in the RAIM algorithm (e.g., by means of a barometric altimeter). The improvement in the fault detection ranges from a factor of 2, for the en route navigation and the terminal phase of flight, up to about 1.4 for the nonprecision approach.

4.3   Differential GPS

As highlighted in Section 4.1, stand-alone GPS (and GLONASS) performance in terms of accuracy and integrity cannot satisfy the requirements of aviation in the most delicate flight phases. Therefore, a supplementary navigation method, called *differential GPS* (DGPS), is used to improve significantly the accuracy and integrity of the stand-alone system [9, 13, 18–25].

DGPS removes correlated errors from receivers (two or more) in visibility of the same satellites: in the basic version, one of the receivers is the monitoring or *reference station* (RS), whose precise position is known. The other receivers (users or *rovers*) have to be in close proximity of the reference station; they have hence to be in line-of-sight (LOS) with it if the reference station communicates with them through a very high frequency (VHF) radio link, which is afflicted by shadowing caused by valleys, buildings, or even trees and by multipath fading. If the reference station uses a medium frequency (MF) radio link, the distances from the rovers can be longer, because an MF wave has a strong ground component able to propagate well beyond the radio horizon. The RS undertakes conventional code-based GPS pseudorange measurements and, due to the knowledge of the position of the RS with a high degree of accuracy, can determine the bias in the measurements. Biases are computed, for each satellite in visibility of the RS, as the difference between the pseudorange measurement and the RS-satellite geometric range. They are affected by errors (due to ionosphere, troposphere, receiver noise, and so forth) induced by the pseudorange measurement process and the receiver clock offset from GPS system time.

In real-time applications, biases (*differential corrections*) are transmitted by the RS to all users in the coverage area. Users exploit these corrections to improve the accuracy of the position solution, thanks to the fact that RS and user receiver pseudorange errors have some common components (due to their proximity), such as those attributable to satellite clock stability. The user receiver can remove these components, achieving more than a factor of 2 improvement in the position error (see Chapter 3).

Other pseudorange error components increasingly differ as the two receivers (user and RS) move farther apart; these components include, for example, ephemeris prediction error, uncorrected satellite perturbations (caused by tidal attractions, solar radiation pressure, solar winds, gas-propellant leakage and others), and also atmospheric errors. For these pseudo-range error components, the correction achieved with DGPS is more accurate the closer the user receiver and the RS are.

Those components that cannot be corrected at the user receiver with DGPS are, for example, multipath, interference, and receiver noise, which are not correlated with the corresponding ones at the RS.

The basic DGPS concept described here is known as *local area DGPS* (LADGPS) or *conventional DGPS* (CDGPS). A basic scheme of the LADGPS concept is shown in Figure 4.1.

Each reference station, after evaluating the pseudorange measurement error at its location, transfers it to the user, for instance through a dedicated data link (mainly VHF links in aviation and medium frequency, MF, links in maritime services).

User receivers closer to the RS (hence experiencing errors having more common components with RS) are able to reach a value of UERE in the order of about 10% of the one related to the nondifferential system.

To give the mathematical basis of the LADGPS operation, suppose that the user receiver and the RS are very close each other, so that only pseudorange error components related to the user segment differ. The reference station must have accurate knowledge of its own position (in ECEF coordinates, see Chapter 2) to allow the user to determine accurately its own position with respect to Earth.

The geometric distance $D_{i,k}$ from the kth reference station to the ith satellite can be obtained from the reported position of the satellite and the RS surveyed position. The RS performs a pseudorange measurement $r_{i,k}$ that can be expressed as:

$$r_{i,k} = D_{i,k} + e_k + o_k \tag{4.4}$$

where e_k is the overall contribution of the space, control, and user segment pseudorange errors and o_k is the RS clock offset from the system time. The RS constructs the differential correction as:

$$\Delta r_{i,k} = r_{i,k} - D_{i,k} \tag{4.5}$$

that is transmitted to the user receiver and compared with the user receiver pseudorange measurement $r_{i,\text{user}}$:

$$r_{i,\text{user}} = D_{i,\text{user}} + e_{\text{user}} + o_{\text{user}} \tag{4.6}$$

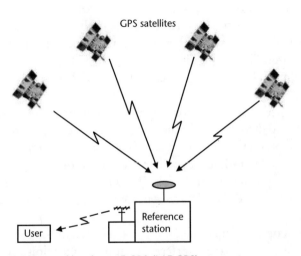

Figure 4.1   Principal scheme of local area DGPS (LADGPS) concept.

Since pseudorange error components related to space and control segments as well as to the signal path are assumed to be the same, due to the proximity of the user receiver and the RS, the corrected pseudorange determined by the user $R_{i,\text{user}}$ can be expressed as the difference between the pseudorange measurement, $r_{i,\text{user}}$, and the differential correction coming from the RS, $\Delta r_{i,k}$:

$$R_{i,\text{user}} = r_{i,\text{user}} - \Delta r_{i,k} = D_{i,\text{user}} + e_{\text{res}} + o_{\text{comb}} \qquad (4.7)$$

where e_{res} is the residual contribution of user segment pseudorange errors and o_{comb} is the combined clock offset. The user position can be determined by the pseudorange measurements from at least four satellites, following one of the techniques highlighted in Chapter 2.

By indicating with τ_j the sample instants when the pseudorange corrections are transmitted by the RS, the transmitted error correction $\Delta r_{i,k}(\tau_j) = r_{i,k}(\tau_j) - D_{i,k}(\tau_j)$ is corrected only at the time τ_j when it was calculated, due to the significant changes between transmissions in the pseudorange error induced by the satellite motion.

Together with $\Delta r_{i,k}(\tau_j)$, the RS provides the user with a pseudorange rate correction $C_{i,k}(\tau_j)$, so that the user adjusts the pseudorange correction to the time t according to the following:

$$\Delta r_{i,k}\left(t_j\right) = \Delta r_{i,k}\left(\tau_j\right) + C_{i,k}\left(\tau_j\right)\left(t - \tau_j\right) \qquad (4.8)$$

The corrected user pseudorange is finally obtained as:

$$R_{i,\text{user}}\left(t\right) = r_{i,\text{user}}\left(t\right) - \Delta r_{i,k}\left(t\right) \qquad (4.9)$$

where the pseudorange error increases with $(t - \tau_j)$, even in the presence of the pseudorange rate correction, due to the satellite radial acceleration.

When the user receiver moves away from the RS, some components of the user receiver pseudorange error that have been considered correlated (i.e., common) with the RS corresponding ones become uncorrelated: these components are those related to satellite perturbations, ephemeris prediction error, and ionospheric and tropospheric delay. This spatial decorrelation introduces differences between the error component determined at the RS and that computed at the user receiver; the differences can be calculated using appropriate expressions and models. Ionosphere and troposphere contributions are shaped through time and space variable parameters that might have large excursion in the range of interest. The models are used to estimate the absolute or residual error components of the pseudorange or corrected pseudorange expressions already given [9, 13].

The region where DGPS corrections can be performed without significant decorrelation can be extended by adding three or more reference stations along the perimeter of the coverage region. In this case, the user achieves a more accurate estimate by a weighed average of the corrections from the stations, where the weights depend only on geometric considerations, with the highest weight given to the closest RS.

From this, it is clear that the loss of accuracy of LADGPS due to spatial decorrelation can be improved with a more sophisticated network of earth stations. This approach is referred to as *wide area DGPS* (WADGPS), where a network of monitoring stations evaluate and continuously update the time and space varying components of the total error over the whole coverage region, rendering the corrections available to users within the coverage area.

WADGPS networks have been implemented in parallel to the development of the U.S. augmentation system that will be described in Section 4.4.1. Private initiatives, as in the case of U.S. offshore oil exploration and seismic survey industries, have largely contributed to the development of private WADGPS services for their applications [9, 26, 27].

The land survey community, together with the offshore industry, has pioneered the DGPS and brought improved high-accuracy measurement techniques, whereas the GIS has exploited DGPS to develop a database of items and corresponding location data.

The WADGPS approach aims at maintaining the LADGPS meter-level accuracy over a large region [9, 13]. It is based on the decomposition and estimation of the pseudorange error components for the whole of the coverage region (or service volume), instead of the close reference position [9, 28, 29]. Therefore, the scalar calculation of the pseudorange error achieved in the LADGPS approach translates, in the WADGPS, into a vectorial evaluation of the error components.

The WADGPS architecture consists of a network composed of at least one MCS, a number of monitor or reference stations, and communications links. Each RS performs the GPS measurements and transfers the data to the MCS. The latter then evaluates the GPS error components, from the known locations of the RSs and the gathered data. The computed error corrections are sent to the user via either telephone or radio/satellite links [13].

The principle of the WADGPS approach and data flow to the user is provided in Figure 4.2, where a four-layer satellite-RS-MCS-user architecture is envisaged. The RS at known locations gathers GPS pseudoranges and navigation data from the visible satellites. The data is then transferred to the MCS, where user differential correction data is evaluated before being transmitted to users. The users apply the corrections to their data to improve the position accuracy. RSs also aid in

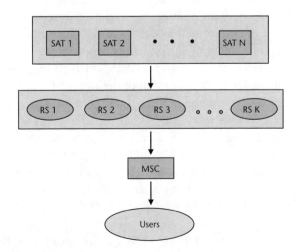

Figure 4.2 The basic WADGPS concept.

determining the satellite ephemeris, atmospheric delay, and discrepancies between the GPS system time and satellite time frames.

In the case of large coverage areas or service volumes, the MCS and the communications system have to cope with high computational and transmission burden to provide and distribute updates to the RSs in a trustworthy and timely manner. Therefore, the principal scheme in Figure 4.2 is to be changed, introducing a further level of stations (e.g., *regional control stations,* or RCSs), which are connected to an inferior number of RSs than the MCS in the architecture of Figure 4.2. Therefore, RCSs can provide timely correction updates and active/standby redundancy to take over the functions of a close RCS, if required. The five-layer architecture is depicted in Figure 4.3. The RCSs operate in conjunction with an MCS, which synchronizes their clocks, coordinates measurements on the satellite, and monitors RCS health. Both the architecture schemes set the synchronization problem of the clocks of all components to the same system time to ensure proper corrections and time-tagging.

The main error sources for WADGPS architecture are satellite ephemeris estimates, reported satellite clock times, and atmospheric delays due to both troposphere and ionosphere. Algorithms and modeling of these errors have being developed to reduce their impact in the overall WADGPS performance [30–33].

A further improvement in the achievable system accuracy, to the submeter level, can be obtained through DGPS techniques that exploit phase information of the GPS satellite signal carrier frequencies [34–38]. The techniques are based on interferometric measurements of satellite carrier frequencies (interferometric GPS, or IGPS). Very high accuracy can be achieved by processing the received satellite signal Doppler frequencies. Doppler frequency shift, which is due to the relative motion between satellite and receiver, on the L1 frequency has to be accounted for by single-frequency receivers, whereas in dual-frequency receivers both L1 and L2 have to be tracked. Error contribution due to multipath can constitute the major obstacle, and ad hoc refinement techniques have been developed to reach the desired centimeter accuracy [34]. Offshore/land surveying and seismic applications

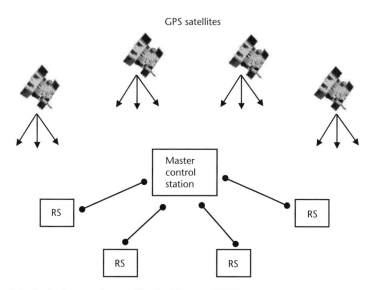

Figure 4.3   Principal scheme of centralized wide area DGPS concept.

have mainly used these *carrier-based techniques*, and the interferometric accuracy is also compliant with aircraft blind landing (i.e., CAT. III precision approach).

In Table 4.2 the GNSS aviation operational performance requirements, expressed in terms of the RNP parameters, are reported [3, 11].

Recent developments in DGPS technologies have led to the development of an innovative technique known as the *virtual reference station* (VRS) network concept (Figure 4.4), where observation data for a nonexisting station are created from the real measurements of a multiple reference station network. Therefore, this new DGPS technique is based on the existence of *networked DGPS* (NDGPS) stations that are connected to a central control station, which performs data correction and modeling. The VRS network concept is the most advanced approach to extend the possible distances between reference stations and user receivers. In addition, it allows one to reduce the number of permanent reference stations necessary to the coverage of a given area [39, 40].

4.4 Satellite-Based Augmentation Systems (SBAS)

The natural evolution of satellite navigation systems is to become a key aid in the connectivity tissue of everyday life, including safety-critical areas.

When "safety-critical" applications are mentioned, the first thought is devoted to aircraft operations. Such an application places a high demand on navigation

Table 4.2   GNSS Aviation Operational Performance Requirements

Operation	Accuracy (95%)	Integrity (1-Risk)	Alert Limit	Time-to-Alert	Continuity (1-Risk)	Availability
Oceanic	12.4 nm	$1-10^{-7}$/hr	12.4 nm	2 min	$1-10^{-5}$/hr	0.99 to 0.99999
En route	2.0 nm	$1-10^{-7}$/hr	2.0 nm	1 min	$1-10^{-5}$/hr	0.99 to 0.99999
Terminal	0.4 nm	$1-10^{-7}$/hr	1.0 nm	30 sec	$1-10^{-5}$/hr	0.99 to 0.99999
NPA	220m	$1-10^{-7}$/hr	0.3 nm	10 sec	$1-10^{-5}$/hr	0.99 to 0.99999
APVI	220m (H) 20m (V)	$1-2\times10^{-7}$/ approach	0.3 nm(H) 50m (V)	10 sec	$1-8\times10^{-6}$/ 15 sec	0.99 to 0.99999
APVII	16m (H) 8m (V)	$1-2\times10^{-7}$/ approach	40m (H) 20m (V)	6 sec	$1-8\times10^{-6}$/ 15 sec	0.99 to 0.99999
CAT. I	16m (H) 4.0–6.0m (V)	$1-2\times10^{-7}$/ approach	40m (H) 10-15m (V)	6 sec	$1-8\times10^{-6}$/ 15 sec	0.99 to 0.99999
CAT. II	6.9m (H) 2.0m (V)	$1-10^{-9}$/15 sec	17.3m (H) 5.3m (V)	1 sec	$1-4\times10^{-6}$/ 15 sec	0.99 to 0.99999
CAT. III	6.2m (H) 2.0m (V)	$1-10^{-9}$/15 sec	15.5m (H) 5.3m (V)	1 sec	$1-2\times10^{-6}$/ 30 sec (H) $1-2\times10^{-6}$/ 15 sec (V)	0.99 to 0.99999

Note: H denotes the horizontal requirement; V denotes the vertical requirement.

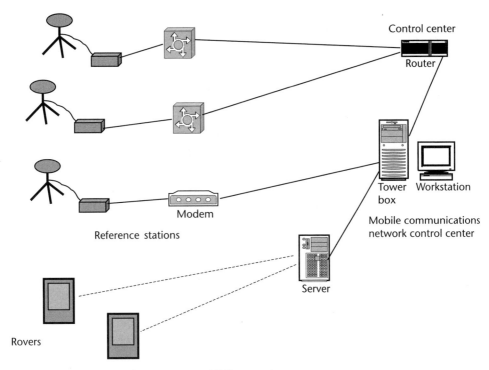

Control center
Router
Tower Workstation
box
Mobile communications
network control center
Modem
Reference stations
Server
Rovers

Figure 4.4   The virtual reference station (VRS) network concept.

systems. The performance of such systems can be specified in terms of the RNP
parameters:

- Integrity monitoring;
- Continuity of service/reliability;
- Time availability;
- Accuracy performance compliance with landing categories I–III.

Section 4.2 discussed an integrity monitoring technique. However, what is
required for aviation is a global system concept able to deal with both technical and
institutional limitations of the current stand-alone systems such as GPS.

Satellite-based augmentation systems (SBAS) represent a first answer to these
demands in terms of technical performance. A further answer (that addresses both
technical and institutional limitations) will be given by second generation satellite
navigation systems, in particular GALILEO and GPS III, which will be discussed in
the next chapters.

The following describes augmentation systems currently under development
and those in the planning stages. Their role in the development of future global
navigation systems can be tightly compared with the 2.5 generation general packet
radio service (GPRS) and enhanced data rates for GSM evolution (EDGE) mobile
systems that represent a soft transition from the key second generation digital system (global system for mobile communications, or GSM) to the third generation
network (universal mobile telecommunications system, or UMTS) [41]. In fact, augmentation systems can be considered as 1.5 generation of GNSS, providing a soft

transition from the GPS/GLONASS systems to the second generation GNSS (GALILEO, GPS III).

4.4.1 WAAS

The *wide area augmentation system* (WAAS) is a safety-critical system composed of a space segment and a ground network designed to support en-route up to CAT. I precision approaches [42] (the accuracy is within the specified vertical requirement, but the integrity requirement is still not fully satisfied [43]). WAAS has been designed to augment GPS to be used as a primary navigation sensor, providing the following three augmented services [44–48]:

- A ranging function that improves availability and reliability;
- Differential GPS corrections, which improve accuracy;
- Integrity monitoring, which improves safety.

The differential corrections at L1 frequency are broadcast from geostationary satellites that "augment" the system space segment; RAIM is exploited to provide integrity when the aircraft is outside the WAAS coverage region or a catastrophic failure occurs.

WAAS distributes GPS integrity and correction data to GPS users and provides a ranging signal that augments GPS. The WAAS signal is broadcast to users from geostationary satellites. Two INMARSAT-3 satellites are currently used as geostationary components. Two further satellites are envisaged for 2005 [49]. The first one is PanAmSat's Galaxy XV, a member of the orbital small- to medium-size STAR spacecraft family. The second satellite, the Telesat ANIK F1R, is a product of the company Astrium. Both launches have been planned for 2005. The current and envisaged WAAS geostationary satellite coverage is displayed in Figure 4.5(a, b), respectively.

The WAAS ranging signal is GPS-like and WAAS receivers are, hence, derived by modifying the GPS receiver. The WAAS signal is at the L1 frequency and is modulated with a spread spectrum code from the same family as the GPS gold C/A codes. Those codes have been selected to not interfere with the GPS signals. Code phase and carrier frequency of the signal will be controlled to provide, by means of WAAS geostationary satellites, additional range measurements to GPS users. The WAAS signal also contains differential corrections and integrity information for all satellites of the space segments (both GPS and geostationary). A WAAS geostationary satellite may broadcast either integrity data only or both integrity and WAAS data. The integrity provides the *use/do not use* data on all visible satellites. The WAAS architectural concept is depicted in Figure 4.6.

The WAAS ground network provides differential corrections and integrity data to the users. The network is composed of *wide area reference stations* (WRS), which are largely dispersed data collection locations where signals received from the GPS satellites and geostationary components are received. The WRS forwards the information to data processing sites, called *wide area master stations* (WMS) or central

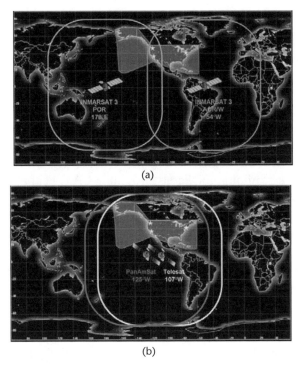

Figure 4.5   WAAS geostationary coverage: (a) current satellites, and (b) future satellites. (Courtesy of the Federal Aviation Administration.)

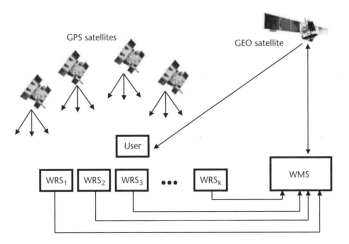

Figure 4.6   WAAS architecture concept.

processing facilities. The WMS processes the raw data to determine integrity, differential corrections, residual errors, and ionosphere delay information for each monitored satellite. Furthermore, they create ephemeris and clock information for the geostationary satellites. Those data are included in the WAAS message that is transferred to the navigation earth stations. The latter transmit the message to the geostationary satellites, which then broadcast the GPS-like signal to users across the United States and Alaska. Besides the two geostationary satellites, the WAAS network currently includes 25 WRS, two WMS, and three uplink navigation earth stations, for use in most flight phases.

The differential corrections together with the improved geometry provided by the WAAS geostationary satellites enhance both horizontal and vertical user accuracy to values ranging from 0.5m to 2m, exceeding the required value of 7.6m, both for horizontal and vertical components [50].

However, the initial technical goals of WAAS had been scaled back; FAA managers stated that WAAS will offer a certified lateral navigation/vertical navigation (LNAV/VNAV) capability (where the decision height is 350 feet with a visibility of one mile), instead of immediately providing CAT. I capabilities (where the decision height must be 200 feet with a visibility of half a mile) [51].

On July 10, 2003, the FAA commissioned WAAS, hence letting WAAS become part of the U.S. National Aerospace System (NAS) [50]. This represented the IOC of WAAS and moved FAA to LNAV/VNAV and localizer precision with vertical (LPV) guidance capabilities using WAAS: both LNAV/VNAV and LPV approaches exploit the accuracy of the WAAS signal to include vertical guidance capability. The FAA has published almost 600 LNAV/VNAV approaches at about 300 U.S. airports that can be used with WAAS. The implementation of LPV approaches further improves precision approach capability to users: LNAV/VNAV vertical accuracy is combined with lateral guidance similar to the typical instrument landing system (ILS). There are currently seven LPV approach locations in the United States, and production will continue until all qualified airports have an LPV approach at each runaway end.

Although WAAS was mainly designed for aviation users, it supports a variety of nonaviation applications, including, for instance, agriculture, surveying, recreation, and surface transportation. In particular, the worldwide agricultural community is facing great challenges, such as how to feed an ever-increasing world population, increase the in-farm profit in order for farmers to maintain their own land, and decrease the impact of modern agriculture on the environment (see Chapter 8). These problems can be solved by innovative tools, such as novel soil sampling techniques and variable rate applicators for accurate placement of fertilizer, chemicals, and seeds. The agriculture industry realized that many of these new tools and techniques require accurate, reliable advanced navigation techniques to provide repeatable field location [52]. Manufacturers and farmers involved in this new precision agriculture industry recognized an opportunity to reduce significantly the equipment cost by incorporating the new WAAS technology into their products.

The WAAS signal has been available for nonsafety-of-life application since 2000. At present, there are millions of WAAS-enabled GPS receivers in use.

The FAA is striving forward to achieve FOC for WAAS in 2007, deploying the full geostationary constellation that will guarantee that each receiver sees at least two geostationary satellites at all times throughout the continental United States and most of Alaska. The FAA is also cooperating with National Canada and SENEAM of Mexico to extend the WAAS coverage area by locating additional WRS in these countries [53]. In addition, the Brazilian government, properly assisted by the FAA, intends to install reference stations and processing centers, based on WAAS technology, to provide SBAS services across Brazil and other South American countries [53].

WAAS is envisaged to continue to evolve to take full advantage of the modernized GPS constellation, including the L5 frequency. In this respect, the FAA is including the L5 frequency in current WAAS geostationary components.

As the WAAS signal is now available, WAAS receivers and WAAS-related procedures are becoming available. WAAS receivers certified for aviation operations are arriving on the market to be available to users. The hectic activity from avionics manufacturers started in late 2002 [54]. An example among WAAS certified receivers (UPSAT, Chelton, Capstone) is shown in Figure 4.7 [54].

There are three functional classes of WAAS receivers [54]:

- *Class Beta*: This receiver generates WAAS-based position and integrity information but does not have its own navigation function. It is certified under technical standard order (TSO)-C145a and is typically used in conjunction with a flight management system.
- *Class Gamma*: This receiver is an integrated beta sensor, navigation function, and database that provides a complete, stand-alone WAAS navigation capability. It is the typical panel-mount receiver used by most general aviation aircraft and is certified under TSO-C146a.
- *Class Delta*: This receiver provides guidance deviations only to a precision final approach (similar to ILS) and consists of a class beta sensor and navigation processor. The requisite database is typically resident in a flight management system and accessed by the class delta receiver. Because a navigation function is included in it, the delta receiver is certified under TSO-C146a.

The basic configurations of the three receiver types—classes beta, gamma, and delta—are depicted in Figure 4.8(a–c), respectively.

Within each functional class, there are four operational classes:

- *Class 1*: Receivers can be used for oceanic, en route, departure, terminal, and nonprecision approach operations.
- *Class 2*: Receivers add the ability to fly LNAV/VNAV approach procedures.
- *Class 3*: Receivers add the ability to fly precision LPV/CAT. I approach procedures.
- *Class 4*: Receivers provide navigation to a precision final approach and do not support other navigation functions.

Figure 4.7   An example of a WAAS receiver: certified UPSAT. (© 2004 Garmin Corp.)

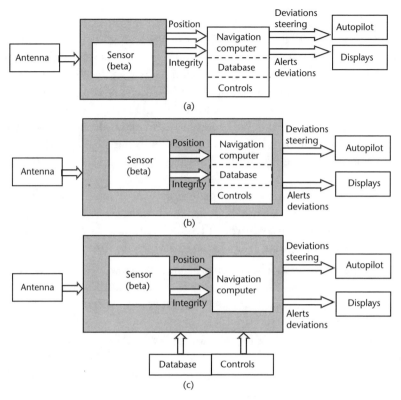

Figure 4.8 Functional class WAAS receiver: (a) class beta configuration, (b) class gamma configuration, and (c) class delta configuration.

4.4.2 EGNOS

EGNOS, the European Geostationary Navigation Overlay Service, is the first European initiative in the satellite navigation field. EGNOS was mentioned for the first time in 1994 [55], in a communication from the European Commission. This was followed by the December 19, 1994, resolution by the Council of European Union [56] to define the terms of the European contribution to the development of a GNSS. EGNOS is a joint activity of the European Commission, European Space Agency, and Eurocontrol. The organizations created a tripartite agreement in which each body has a specific task [57]:

- The European Space Agency is responsible for the technical development of EGNOS and its operation for testing and technical validation purposes.
- Eurocontrol provides the civil-aviation user requirements and validates the resulting system in light of these requirements.
- The European Commission contributes to the consolidation of the requirements of all users and to the validation of the system in light of such requirements, in particular in the framework of its trans-European networks and research and development actions. It also provides for the establishment of EGNOS by taking all appropriate measures, including the leasing of geostationary transponders.

The EGNOS program is an integral part of the European satellite radio-navigation policy. It is currently under the control of GALILEO JU (see Chapter 5). The aim of EGNOS, like the other SBAS services, is to provide complementary information to the GPS and GLONASS signals to improve the RNP parameters (see Chapter 2) [58–63].

The overall system architecture is divided into three segments (Figure 4.9) [62]:

- Space segment;
- Ground segment;
- User segment.

The space segment consists of three geostationary Earth orbiting (GEO) satellites that provide triple coverage over Europe, the Mediterranean, and Africa. EGNOS currently foresees the use of two INMARSAT-3 satellites and the ARTEMIS (Advanced Relay Technology Mission) satellite (positioned 21.5°E and PRN code is 124), as shown in Figure 4.10. The INMARSAT satellites are, respectively, AOR-E 3F2 (positioned 15.5°W and PRN code is 120) and IOR-W 3F5 (positioned 25°E and PRN code is 126). The IOR 3F1 INMARSAT satellite (positioned 64°E and PRN code is 131) in use for EGNOS System Test Bed (ESTB) has been discontinued. The next generation of INMARSAT-4 satellites will embark a navigation payload; one of those satellites will be located in the existing INMARSAT IOR orbit location. The navigation payload is a bent-pipe transponder enabling the messages uploaded to the satellites to be broadcast to users using a GPS-like signal.

The ground segment includes the following elements:

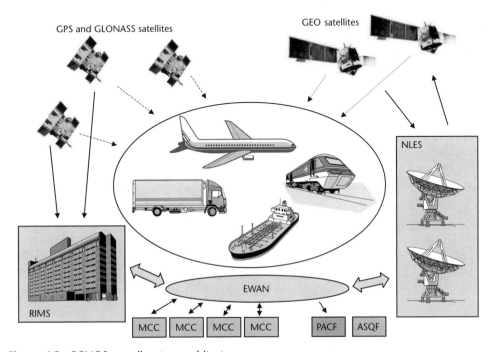

Figure 4.9 EGNOS overall system architecture.

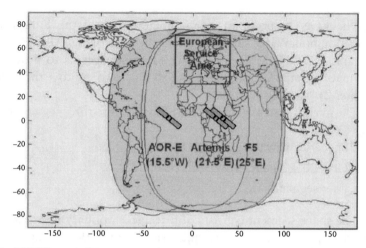

Figure 4.10 EGNOS geostationary coverage

- Four *mission control centers* (MCC) that include a *central control facility* (CCF) and a *central processing facility* (CPF);
- Thirty-four *ranging and integrity monitoring stations* (RIMS);
- Six *navigation land earth stations* (NLES);
- The *application specific qualification facility* (ASQF);
- The *performance assessment and system checkout facility* (PACF);
- The *EGNOS wide area communication network* (EWAN).

These elements are spread over European territory and surrounding continents, as shown in Figure 4.11. The non-European RIMS are spatially sited in line with international cooperation strategy, as discussed in Chapter 5.

The RIMSs are reference stations and they have several main functions at system level [64]; they are provided with atomic clocks. The EGNOS timing requirements are based on a bespoke time reference, the EGNOS network time (ENT). ENT corrections are broadcast to the RIMSs with an accuracy of 1.5 ns and the time reference station for EGNOS is located in the *Observatoire de Paris*. The RIMS measure satellite pseudoranges measurements (code and phase) from GPS/GLONASS and SBAS GEO satellites signals. The raw measurements are transmitted to the CPF, which determines the wide area differential (WAD) corrections and ensures the integrity of the EGNOS system for users.

The RIMS can have two or three channels feeding the CPF called (channel A/B and channel A/B/C). Each channel is separated from the others (and developed by different main contractors) and transmits data to the EWAN. Channel A is used to transmit raw data for the computation of the differential corrections; Channel B is linked with CPF check chain for comparison and integrity monitoring purposes. Channel C is used for dedicated integrity function (*satellite failure detection*, or SFD).

The RIMS continuously checks the GPS and GLONASS signal to detect failures onboard the satellite, which imply errors in the measured satellite signal correlation function. In case an error is detected, the RIMS raises a flag and sends it to the CPF, to generate the DON'T USE flag for users. The local anomalies, such as local

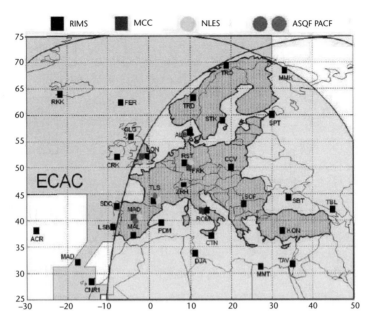

Figure 4.11 EGNOS ground segment architecture.

interference or multipath, are computed at the RIMS site. The RIMS can also be remotely controlled by the central control facility that is included in each MCC. System architecture foresees four MCCs, located in Torrejon, Spain, Gatwick, United Kingdom, Langen, Germany, and Ciampino, Italy; only one of the MCCs is active and operational, whereas the remaining are hot spares to be activated if a problem occurs.

The MCC transmits the EGNOS messages to the six NLES. Each NLES broadcasts the EGNOS messages to users via three geostationary satellites. There is a spare NLES for each GEO satellite. The ASQF and PACF are located, respectively, in Torrejon, Spain, and Toulouse, France. The PACF is a support facility, whose tasks include technical coordination, performance analysis, and system configuration management. The ASQF is a support facility as well and provides a platform for the validation and certification of EGNOS applications.

The EGNOS user segment is composed of a GPS and/or GLONASS receiver and EGNOS receiver. The two receivers are usually embedded in the same user terminal. The receiver can process the message that is scheduled in a 6-second duty cycle. The integrity time to alarm is, hence, limited to the duty cycle time. The EGNOS message includes slow correction (more slowly changing errors, such as long-term satellite clock drift, long-term orbital error correction, and ionosphere delay corrections) and fast correction (rapidly changing errors, such as satellite clock errors) in the same frame, as shown in Figure 4.12 [60].

In the context of the development of effective interfaces between the navigation and the communications component, it is worth mentioning the Signal In Space through the Internet (SISNET). This is a means to integrate satellite navigation, specifically the EGNOS system, and the Internet. It cannot be considered a simple add-on of EGNOS, as it enables a number of services and applications that the stand-alone EGNOS is unable to provide.

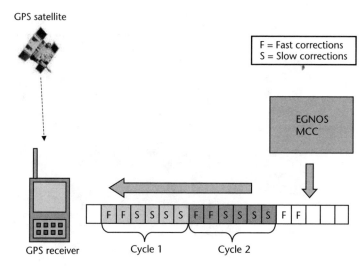

Figure 4.12 EGNOS message composition.

The main aim of EGNOS was to support the safety of life applications, specifically aviation requirements. EGNOS receivers need LOS visibility to receive the signals from the geostationary satellites and this is not an issue for aircrafts. The main aim of SISNET is to bring EGNOS messages to a wide set of terrestrial applications.

The SISNET system architecture is shown in Figure 4.13 [65]. The communications protocol is called Data Server to Data Client (DS2DC) and is totally based on the well-known Transmission Control Protocol/Internet Protocol (TCP/IP), which characterizes the Internet. The user can download EGNOS messages offline by a Web interface or in real time by a TCP/IP protocol; its data rate is very low (about 1 kbit/s). This means that almost every access to the Internet is enough to receive real-time SISNET messages. GPS receivers can fix an approximate position, even if the user is in an urban environment, with the help of other sensors or the support of a

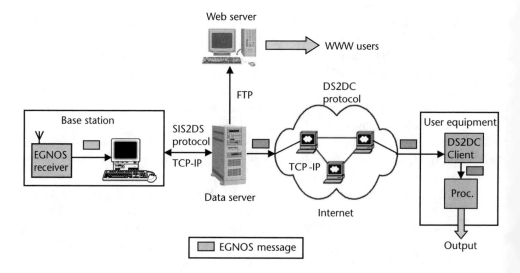

Figure 4.13 SISNET system architecture.

wireless network (see Chapter 9). SISNET also allows an improvement in the RNP parameters, in difficult environments where geostationary satellites are masked, and the integrity certification of the computed position.

ESA is currently financing new applications of SISNET, including the development of a navigation user terminal that will help blind people navigate in urban environments and a new system of European vehicular remote tolling service.

The future development of the EGNOS project envisages an increase of the number of RIMSs and geostationary satellites. The European Commission is going to create international agreements to bring the EGNOS system and the RIMSs to the Mediterranean area and Africa. Nevertheless INMARSAT-4 satellites will include a navigation payload for EGNOS and the other SBAS systems and they will expand and improve the EGNOS signals. Due to its mentioned integration in the GALILEO Programme and control by the JU, EGNOS is clearly the forerunner of the GALILEO system [57].

4.4.3   Extension of SBAS Systems

As outlined in the previous pages, satellite-based augmentation is a fascinating engineering approach that allows the exploitation of existing satellites to create a performance-enhanced combined system that anticipates some of the features of next generation systems. In this respect, satellite-based augmentation is becoming a worldwide necessity, if we pose seriously the difficult question:

> When you get a GPS navigation signal, how do you know you can trust it?
> —Laurent Gauthier

Certainly, WAAS and EGNOS have paved the way for a worldwide diffusion of SBAS.

C-WAAS is the Canadian extension of the U.S. WAAS system. It is based on a network of reference stations located in the Canadian territory and linked to the FAA master stations in the United States.

Japan is developing an SBAS based on its Multi-function Transport Satellites (MTSat), called the MTSat Satellite Augmentation System (MSAS). MSAS is being developed by the Civil Aviation Bureau, Ministry of Land, Infrastructure and Transport, Government of Japan [66, 67]. The launch of the first geostationary satellite of the MTSat family (MTSat-1) failed in 1999. As a consequence, the MSAS program has been delayed and its operational phase is now foreseen for 2005. The replacement satellite MTSat-1R is scheduled to be launched in 2005 and a second geostationary satellite (MTSat-2) should follow with a launch by 2008.

India is developing its own SBAS called GAGAN (GPS and GEO Augmented Navigation) and it is under the control of the Indian Space Research Organisation and the Airports Authority of India. Operational capability is planned for 2007 [68–70].

The People's Republic of China launched two satellites in 2000 called Beidou 1A, located at 140°E, and Beidou 1B, located at 80°E, and a third one, Beidou 1C, located at 110°E, in 2003. This represented the first Chinese experimental navigation technology satellites. The Chinese SBAS strategy is encapsulated in the Satellite Navigation Augmentation System (SNAS).

The complex scenario of the worldwide SBAS program is summarized in Figure 4.14. Interoperability at the receiver level is the compliance with DO-229C standard [71–73] and it is guaranteed (at the moment) by EGNOS, MSAS, WAAS, and C-WAAS.

4.5 Ground-Based Augmentation Systems (GBAS)

The concept of enhancing basic GPS performance through terrestrial aids was conceived some years ago. Pseudolites (PL), or integrity beacons, are low-power ground-based transmitters that can be configured to emit GPS-like signals for improving accuracy, integrity, and availability of GPS [13, 74–79]. The improvement in accuracy can be obtained due to the enhancement in local geometry, measured according to a lower VDOP, that is of key importance in aircraft precision approach and landing. Accuracy and integrity improvement can also be achieved by exploiting the pseudolite integral data link to support DGPS operations and integrity warnings broadcast. Furthermore, PLs provide additional ranging sources that augment the GPS constellation, hence enhancing availability. The use of PL implies a careful evaluation of some induced effects, such as the near-far problem that a user receiver may experience, depending on the signal level encountered as the distance to the PL varies.

The PLs typically operate at the L1 frequency and are modulated with an unused pseudorandom code, in order not to be confused with a satellite. A second GPS antenna mounted on the aircraft is used to acquire the PL signal. The presence of two PLs, which are generally placed within several miles of a runaway threshold along the nominal approach path, reduces the requirement of visible satellites for DGPS operations to four. Centimeter-level position accuracy can be achieved [75, 77–79].

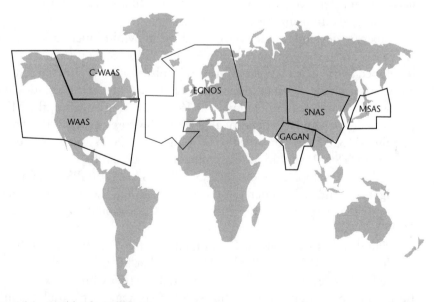

Figure 4.14 Worldwide SBAS systems.

The performance enhancement concept based on ground support is being further developed by the FAA, which is deploying a *ground based augmentation system* (GBAS) named *local area augmentation system* (LAAS) [80–84].

LAAS has been conceived to provide on a local basis differential corrections and integrity monitoring, resulting in an accuracy performance in the order of 1m. Obviously the better accuracy, with respect to the WAAS figures, is counterbalanced by the reduced extension of the operation area (30–40 km). An LAAS architecture basically consists of:

- A differential *reference station;*
- A *monitoring station*, which receives signals from the satellites and the reference station and, on that basis, issues the *go-not go* message: the latter informs, by means of a proper message, typically at VHF, the users about the usability or not of the system;
- Eventual *pseudolites* positioned around the runaway to augment the GNSS (in particular GPS) measurements available to the user.

The FAA is giving large emphasis to the LAAS program: CAT. I LAAS Ground Facility (LGF) contract is underway with Honeywell and this is illuminating the future LAAS CAT. II/CAT. III program [85]. The latter is proceeding in two phases: phase one is devoted to determine the architecture that will be used for CAT. II/CAT. III precision approaches. Phase two will allow the exploitation of dual frequencies, including the new GPS L5 civil frequency or the future GALILEO E5a or E5b frequencies, together with the phase one L1 frequency.

The complex LAAS development process has been started at six U.S. airports: the Juneau, Alaska, Phoenix, Chicago, Memphis, Houston, and Seattle airports. In particular, LAAS application in a very difficult airport, Alaska's capital, Juneau, is very interesting. The place experiences more than 320 days of precipitation a year, intense clouds, and is surrounded by heavy mountains [86]. The FAA LAAS program hopes to play a significant role in rendering Juneau arrivals and departures more accessible, more dependable, and safer: If LAAS can improve approaches in Juneau, it can certainly do it elsewhere.

LAAS users have identified some major benefits of the LAAS approach with respect to ILS, including the ability of supporting complex terminal area procedures (including missed approaches), implementing multiple, segmented or variable glide-paths, supporting extended arrival procedures, reducing air traffic controller workload, supporting RNP operations, and supporting adjacent airport operations [87].

It is expected that LAAS will have reduced siting constraints, when compared with existing terrestrial navigation aids (ILS, VOR). The LAAS architecture envisages the dislocation on-ground of VHF data broadcast (VDB) antennas, reference receiver antenna (RRA) groups, and LAAS LGF shelter, that includes the processing and transmitting equipment. LAAS equipment can be potentially sited on the airport's existing topography without alterations to existing surroundings or acquisition of further estate [88]. Furthermore, LAAS system modularity allows the addition of components aimed at mitigating site-specific issues that would otherwise impact system performance.

Interoperability of ground- and satellite-based augmentation components represents a key issue for the effectiveness of 2.5G navigation systems. Figure 4.15 highlights the interoperability potential of WAAS/LAAS in aviation: WAAS optimal exploitation concerns en route oceanic and domestic flight phases, WAAS/LAAS combined or alternate operations cover terminal and approach flight phases, and LAAS-only operations pertain to surface services.

4.6 Aircraft-Based Augmentation Systems (ABAS)

The augmentation and/or integration of information obtained through a GNSS with onboard aircraft information leads to the concept of *aircraft-based augmentation systems* (ABAS). Section 4.2 discussed the RAIM method: it is the basic ABAS example. Its performance can be improved applying the concept of *aircraft autonomous integrity monitoring* (AAIM), which uses data coming from both satellite navigation systems and other navigation data sources, such as DME, VOR, LORAN-C, Omega, precise clocks, INS, and aerodynamic and thermodynamic sensors. Chapter 9 deals with the integration of the information achieved by these systems with that obtained by satellite navigation systems. Here, the authors wish to provide figures of the ABAS system performance with respect to RNP parameters.

The integration of GPS with a barometer and a precise clock is very simple; it had been demonstrated that barometer aid increases the mean availability of autonomous integrity monitoring over GPS alone to about 100% for oceanic and en route flight phases, whereas NPA availability can reach about 90%, which is below the requirement for this flight phase [89].

If the GPS navigation data have to be integrated with those coming from radio navigation systems, the complexity is greater; in fact, various problems have to be

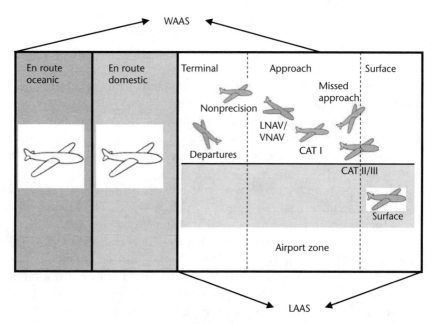

Figure 4.15 Integration between WAAS and LAAS in aviation.

solved, for example, the position determination of the ground radio transmitter stations must be performed in WGS-84 coordinates or the identification and modeling of the errors and failure modes must be realized.

Studies on system integrity about combining GPS data with LORAN-C for non-precision approach have demonstrated an improvement on GPS from 97.335% to 99.982% for failure detection, whereas for FDI the values are 46.2% and 97.72% [90]. Therefore, GPS augmented with LORAN-C does not yet meet the navigation integrity requirement for NPA.

GPS and INS have complementary characteristics, hence rendering their integration effective, also in terms of the RNP parameters that are satisfied up to the NPA approach flight phase. In particular, it is reported that a 100% availability of fault detection and exclusion for NPA is reached [91].

References

[1] *GPS Risk Assessment Study*, Report, The Johns Hopkins University, VS-99-007, M801, January 1999.

[2] Jansen, A. J., "Real-Time Ionospheric Tomography Using Terrestrial GPS," *ION GPS-98*, Nashville, TN, September 1998, pp. 717–727.

[3] Washington, Y., et al., "GPS Integrity and Potential Impact on Aviation Safety," *Journal of Navigation*, Vol. 56, No. 1, Cambridge University Press, January 2003, pp. 51–65.

[4] Barker, B., and Huser S., "Protect Yourself! Navigation Payload Anomalies and the Importance of Adhering to ICD-GPS-200," *Proc. of ION GPS-98*, Nashville, TN, 1998.

[5] Cobb, H. S., et al., "Observed GPS Signal Continuity Interruptions," *Proc. of ION GPS-95*, Palm Springs, CA, 1995.

[6] Walsh, D., and P. Daly, *Definition and Characterisation of Known and Expected GPS Anomaly Events*, Final Report to the U.K. CAA (Safety Regulation Group), 2000.

[7] Pullen, S., Xie, G., and Enge, P., "Soft Failure Diagnosis and Exclusion for GBAS Ground Facilities," *Proc. of RIN NAV 2001*, London, England, 2001.

[8] "Global Positioning System Standard Positioning Service Performance Standard," U.S. Department of Defense, October 2001.

[9] Kaplan, E. D., *Understanding GPS: Principles and Applications*, Norwood, MA: Artech House, 1996.

[10] Frei, E., and G. Beutler, "Rapid Static Positioning Based on Fast Ambiguity Resolution Approach 'FARA': Theory and First Results," *Manuscript Geodetica*, Vol. 15, 1990.

[11] "Validated ICAO GNSS Standards and Recommended Practices (SARPS)," International Civil Aviation Organization, November 2000.

[12] Niesner, P. D., and Johannsen, R. "Ten Million Datapoints from TSO-Approved GPS Receivers: Results and Analysis and Applications to Design and Use in Aviation," *Navigation*, Vol. 47, No. 1, 2000, pp. 43–50.

[13] Parkinson, B. W., and J. J. Spilker Jr., (eds.), "Global Positioning System: Theory and Applications," *Progress in Astronautics and Aeronautics*, Vol. 164, American Institute of Aeronautics and Astronautics, 1996.

[14] Van Dyke, K. L., *Analysis of Worldwide RAIM Availability for Supplemental GPS Navigation*, DOT-VNTSC-FA360-PM-93-4, May 1993.

[15] Brown, R. G., et al., *ARP Fault Detection and Isolation: Method and Results*, DOT-VNTSC-FA460-PM-93-21, December 1993.

[16] Van Dyke, K. L., "RAIM Availability for Supplemental GPS Navigation," *Navigation*, Vol. 39, No. 4, Winter 1992–93, pp. 429–443.

[17] Van Dyke, K. L., "Fault Detection and Exclusion Performance Using GPS and GLONASS," *Proc. of the ION National Technical Meeting*, Anaheim, CA, January 18–20, 1995, pp. 241–250.

[18] Gehue, H., and W. Hewerdine, "Use of DGPS Corrections with Low Power GPS Receivers in a Post SA Environment," *Proc. of IEEE Aerospace Conference*, Vol. 3, Big Sky, MT, March 10–17, 2001, pp. 3/1303–3/1308.

[19] Vickery, J. L., and R. L King, "An Intelligent Differencing GPS Algorithm and Method for Remote Sensing," *IEEE International Geoscience and Remote Sensing Symposium*, Vol. 2, June 24–28, 2002, pp. 1281–1283.

[20] Raquet, J. F., "Multiple GPS Receiver Multipath Mitigation Technique," *IEEE Proc. Radar, Sonar and Navigation*, Vol. 149, No. 4, August 2002, pp. 195–201.

[21] Ito, M., K. Kobayashi, and K. Watanabe, "Study on the Method to Control the Autonomous Vehicle by Using DGPS," *Proc. of the 41st SICE Annual Conference*, Vol. 4, Osaka, Japan, August 5–7, 2002, pp. 2382–2384.

[22] Chen, G., and D. A. Grejner-Brzezinska, "Land-Vehicle Navigation Using Multiple Model Carrier Phase DGPS/INS," *Proc. of the 2001 American Control Conference*, Vol. 3, Arlington, VA, June 25–27, 2001, pp. 2327–2332.

[23] Soares, M. G., B. Malheiro, and F. J. Restivo, "An Internet DGPS Service for Precise Outdoor Navigation," *IEEE Conference Emerging Technologies and Factory Automation*, Vol. 1, September 16–19, 2003, pp. 512–518.

[24] Shuxin, C., W. Yongsheng, and C. Fei, "A Study of Differential GPS Positioning Accuracy," *3rd International Conference on Microwave and Millimeter Wave Technology 2002*, August 17–19, 2002, pp. 361–364.

[25] Farrell, J., and T. Givargis, "Differential GPS Reference Station Algorithm-Design and Analysis," *IEEE Trans. on Control Systems Technology*, Vol. 8, No. 3, May 2000, pp. 519–531.

[26] Lapucha, D., and M. Huff, "Multi-Site Real-Time DGPS System Using Starfix Link: Operational Results," *Proc. ION GPS-92*, Albuquerque, NM, September 1992, pp. 581–588.

[27] Mack, G., G. Johnston, and M. Barnes, "Skyfix-The Worldwide Differential GPS System," *Proc. DNSN-93*, Amsterdam, the Netherlands, April 1993.

[28] Brown, A., "Extended Differential GPS," *Navigation*, Vol. 36, No. 3, Fall 1989.

[29] Leick, A., *GPS Satellite Surveying*, New York: John Wiley & Sons, 1990.

[30] Wallenhof, H. B., H. Lichtenegger, and J. Collins, *GPS Theory and Practice*, New York: Springer-Verlag, 1992.

[31] Loomis, P., L. Sheiblatt, and T. Mueller, "Differential GPS Network Design," *Proc. ION GPS-9*, Albuquerque, NM, September 1991, pp. 511–530.

[32] Kee, C., B. W. Parkinson, and P. Axelard, "Wide Area Differential GPS," *Navigation*, Vol. 38, No. 2, Summer 1991, pp. 123–145.

[33] Ashkenazi, V., C. J. Hill, and J. Nagel, "Wide Area Differential GPS: A Performance Study," *Proc. 5th Intern. ION GPS-92*, Albuquerque, NM, September 1992, pp. 589–598.

[34] Van Grass, F., and M. Braasch, "GPS Interferometric Attitude and Heading Determination: Initial Flight Results," *Navigation*, Vol. 38, No. 4, Winter 1991–1992, pp. 297–316.

[35] Counselman, C., and S. Gourewitch, "Miniature Interferometer Terminals for Earth Surveying: Ambiguity and Multipath with Global Positioning System," *IEEE Trans. on Geoscience and Remote Sensing*, Vol. GE-19, No. 4, October 1981.

[36] Greenspan, R.L., et al., "Accuracy of Relative Positioning by Interferometry with Reconstructed Carrier, GPS Experimental Results," *Proc. 3rd Intern. Geodetic Symposium on Satellite Doppler Positioning*, Las Cruces, NM, February 1982.

[37] Hatch, R., "The Synergism of GPS Code and Carrier Measurements," *Proc. ION GPS-90*, Colorado Springs, CO, September 1990.

[38] Abidin, H., "Extrawidelaning for 'On the Fly' Ambiguity Resolution; Simulation of Multipath Effects," *Proc. ION GPS-93*, Salt Lake City, UT, September 1993, pp. 831–840.

[39] Retscher, G., "Accuracy Performance of Virtual Reference Station (VRS) Networks," *Journal of Global Positioning Systems*, Vol. 1, No. 1, 2002, pp. 40–47.

[40] Trimble (Terrasat), "Introducing the Concept of Virtual Reference Stations into Real-Time Positioning," *Technical Information*, 2001.

[41] Prasad, R., and M. Ruggieri, *Technology Trends in Wireless Communications*, Norwood, MA: Artech House, 2003.

[42] Gustafson, D. M., A. E. Smith, and R. Cassell (Rannoch Corporation), "Cost Benefit Analysis of the Combined LAAS/WAAS System," *AIAA 17th Annual Digital Avionics Systems Conference*, 1998.

[43] "GPS Integrity and Potential Impact on Aviation Safety," Civil Aviation Authority (Safety Regulation Group), CAA Paper 2003/9, April 2004; http://www.caa.co.uk.

[44] Enge P., et al., "Wide Area Augmentation for the GPS," *Proc. of the IEEE*, Vol. 84, No. 8, August 1996.

[45] Di Esposti R., H. Bazak, and M. Whelan, "WAAS Geostationary Communication Segment (GCS) Requirements Analysis," *Proc. IEEE PLANS*, April 2002, pp. 283–290.

[46] Hu, Z., et al., "The Quality of Service Evaluating Tool for WAAS," *Proc. IEEE Communications, Circuits and Systems Int. Conf.*, Vol. 2, July 2002, pp. 1571–1575.

[47] Lo, S. C., et al., "WAAS Performance in the 2001 Alaska Flight Trials of the High Speed LORAN Data Channel," *Proc. IEEE PLANS*, April 2002, pp. 328–335.

[48] Wright, M., "Human Factors and Operations Issues in GPS and WAAS Sensor Approvals: A Review and Comparison of FAA and RTCA Documents," U.S. Dept. of Transportation, Federal Aviation Administration, Office of Aviation Research, U.S. Dept. of Transportation, Research and Special Programs Administration, John A. Volpe National Transportation Systems Center, Available to the Public Through the National Technical Information Service, 1997.

[49] Tisdale, B., "Additional Geostationary Satellites for WAAS," *SatNav News*, Vol. 21, November 2003, pp. 5–6.

[50] Davis, M. A., "WAAS Is Commissioned," *SatNav News*, Vol. 21, November 2003, pp.1–2.

[51] Divis, D. A., "Augmented GPS: All in the Family," *GPS World*, April 1, 2003.

[52] Fly, E., "WAAS Helping to Solve Worldwide Agricultural Problems," *SatNav News*, Vol. 20, June 2003, pp. 6–7.

[53] Sigler, E., "WAAS International Expansion," *SatNav News*, Vol. 20, June 2003, p. 2.

[54] Beal, B., "WAAS Receivers," *SatNav News*, Vol. 20, June 2003, pp. 8–10, and Vol. 19, April 2003, pp. 6–7.

[55] Communication of the European Commission 248 of June 14, 1994.

[56] Council Resolution of December 19, 1994, on the European Contribution to the Development of a Global Navigation Satellite System (GNSS) (94/C 379/02).

[57] Communication From the Commission to the European Parliament and the Council on Integration of the EGNOS Programme in the GALILEO Programme, Brussels, 19.3.2003, COM(2003) 123 final.

[58] Bretz, E. A., "Precision Navigation in European Skies," *IEEE Spectrum*, Vol. 40, No. 9, September 2003, p. 16.

[59] Tossaint, M. M. M., et al., "Verification Techniques for the Assessment of SBAS Integrity Performances: A Detailed Analysis Using Both ESTB and WAAS Broadcast Signals," *GNSS 2003*, Grat, Austria, May 2003.

[60] Solari, G., J. Ventura-Traveset, and C. Montefusco, "The Transition from ESTB to EGNOS: Managing User Expectation," *GNSS 2003*, Grat, Austria, May 2003.

[61] Comby, D., R. Farnworth, and C. Macabiau, "EGNOS Vertical Protection Level Assessment," *ION GPS 2003*, Portland, OR, 2003.

[62] Ventura-Traveset, J., P. Michel, and L. Gauthier "Architecture, Mission and Signal Processing Aspects of the EGNOS System: The First European Implementation of GNSS," *DSP 2001*, Boston, MA, October 2001.

[63]   Oosterlinck, R., and L. Gauthier, "EGNOS: The First European Implementation of GNSS - Project Status," *IAF 2001*, October 2001.

[64]   Brocard, D., T. Maier, and C. Busquet, "EGNOS Ranging and Integrity Monitoring Stations (RIMS)," *GNSS 2000*, Edinburgh, United Kingdom, May 2000.

[65]   Toran-Martin, F., J. Ventura-Traveset, and J. C. de Mateo, "Internet-Based Satellite Navigation Receivers Using EGNOS: The ESA SISNET Project," *NAVITEC 2001*, Noordwijk, the Netherlands, December 10–12, 2001.

[66]   Ueno, M., et al., "Assessment of Atmospheric Delay Correction Models for the Japanese MSAS," *ION GPS 2001*, Salt Lake City, UT, September 2001.

[67]   Shimamura, A., "MSAS (MTSAT Satellite-Based Augmentation System) Project Status," *Proc. GNSS 98*, Toulouse, France, October 1998.

[68]   *Report of Committee B to the Conference on Agenda Item 6, Proc. 11th Air Navigation Conference*, Montreal, Canada, October 2003.

[69]   "GAGAN: The FAA and India Take Initial Steps," *SatNav News*, Vol. 16, June 2002.

[70]   Singh, S., "India to Use Global Positioning in Civil Air Navigation System," *IEEE Spectrum*, Vol. 39, No. 4, April 2002, pp. 27–28.

[71]   "Minimum Operational Performance Standards for Global Positioning System/Wide Area Augmentation System Airborne Equipment," RTCA DO-229C.

[72]   Ventura-Traveset, J., et al., "Interoperability Between EGNOS and WAAS: Tests Using ESTB and NSTB," *GNSS 2000*, Edinburgh, United Kingdom, May 2000.

[73]   Ventura-Traveset, J. et al., "STID: SBAS Technical Interface Document (STID) for Interoperability," EGNOS, WAAAS, Canadian WAAS and MSAS Jointly Produced Document at Interoperability Working Group, March 22, 1999.

[74]   Van Dierendonck, A. J., B. D. Elrod, and W. C. Melton, "Improving the Integrity Availability and Accuracy of GPS Using Pseudolites," *Proc. NAV 89*, London, England, October 1989, paper 32.

[75]   Schuchman, L., B. D. Elrod, and A. J. Van Dierendonck, "Applicability of an Augmented GPS for Navigation in the National Airspace System," *Proc. IEEE*, Vol. 77, No. 11, November 1989, pp. 1709–1727.

[76]   Van Dierendonck, A. J., "The Role of Pseudolites in the Implementation of Differential GPS," *Proc. PLANS 90*, Las Vegas, NV, March 1990.

[77]   Elrod, B. D., and A. J. Van Dierendonck, "Testing and Evaluation of GPS Augmented With Pseudolites for Precision Landing Applications," *Proc. DSNS 93*, Amsterdam, the Netherlands, March 1993.

[78]   Elrod, B. D., K. J. Barltrop, and A. J. Van Dierendonck, "Testing of GPS Augmented With Pseudolites for Precision Approach Applications," *Proc. ION GPS-94*, Salt Lake City, UT, September 1994.

[79]   Cohen, C. E., et al., "Real Time Flight Testing Using Integrity Beacons for GPS Category III Precision Landing," *Navigation*, Vol. 41, No. 2, Summer 1994.

[80]   Perrin, E., B. Tiemeyer, and E. Smith, "GBAS CAT-I Safety Assessment First Achievements," *Proc. of the 21st Digital Avionics Systems Conference*, Vol. 1, Irvine, CA, October 2002, pp. 3D5-1–3D5-12.

[81]   EUROCONTROL/DFS: "GBAS CAT. I, Safety Assessment - Concept of Operations," 1st Draft, October 23, 2002.

[82]   EUROCONTROL: "Category-I (CAT-I) Ground-Based Augmentation System, (GBAS) Pre-Concept Functional Hazard Assessment," Vol. 1.0, June 28, 2002.

[83]   Murphy, T., and R. Hartman, "The Use of GBAS Ground Facilities in a Regional Network," *Proc. IEEE PLANS*, March 2000, pp. 514–521.

[84]   Macabiau, C., and E. Chatre, "Impact of Evil Waveforms on GBAS Performance," *Proc. IEEE PLANS*, March 2000, pp. 22–29.

[85]   Lay, R., N. G. Mathur, and R. Shetty, "LAAS CAT. II/III Update," *SatNav News*, Vol. 21, November 2003, pp. 4–5, and Vol. 19, April 2003, pp. 7–8

[86]   Montley, C., "LAAS in Alaska," *SatNav News*, Vol. 21, November 2003, pp. 3–4.

[87]   Beal, B., "Complex LAAS Procedures," *SatNav News*, Vol. 19, April 2003, pp. 8–9.

[88]   Clark, G., "LAAS Siting," *SatNav News*, Vol. 19, April 2003, pp. 11–12.

[89]   Lee, Y. C., "Analysis of RAIM Function Availability of GPS Augmented with Barometric Altimeter Aiding and Clock Coasting," *ION GPS-92*, Albuquerque, NM, September 1992.

[90]   Weitzen, J. A., Carroll, J. V, and Rome, H. J. "RAIM Availability of GPS Augmented with LORAN-C and Barometric Altimeter for Use in Non-Precision Approach," *Navigation*, Vol. 43, No. 1, 1996.

[91]   Diesel, J., and Dunn, G., "GPS/IRS AIME: Certification for Sole Means and Solution to RF Interference," *ION GPS-96*, Kansas City, MO, 1996.

CHAPTER 5

GALILEO

5.1 Introduction

In February 2004, the United States and the European Commission, joined by the European Union member states, had a successful round of negotiations in Brussels, where they reached an important agreement on the GPS/GALILEO cooperation [1]. This represents a key step for the GALILEO program, which recognizes the United States and Europe as equal partners in the navigation field and creates optimal conditions for the development of the European navigation system, fully independent, compatible, and interoperable with GPS. This agreement will allow users to exploit in a complementary way both systems with the same receiver, hence creating the world standard of radio navigation by satellite [1, 2]. The agreement establishes clear rules for both parties to jointly or individually continuously improve the performance of their respective systems, for the benefit of all users worldwide.

In February 2004, another important event took place: the GALILEO JU completed the first phase in the process to select the future concession holder of the European satellite navigation system. The second phase will continue with the consortia of companies on the short list. The selection results thus far are encouraging, as they demonstrate the willingness of the private sector to invest in GALILEO [3].

These events provide a good picture of GALILEO: a challenging idea that is coming to reality.

As highlighted in Chapter 1, GALILEO is the European radio navigation program that was launched by the European Commission and developed jointly with the European Space Agency [4–9]. It opens the door to a revolution comparable to that generated by mobile phones. In addition, GALILEO promises a new generation of universal services in various sectors, such as transport, telecommunications, agriculture, and fisheries.

The GALILEO program will be managed and controlled by civilians and will offer a guarantee of quality and continuity that is essential for many sensitive applications. These important aspects will be highlighted in this chapter.

The different organization of this chapter with respect to Chapter 3, dedicated to the GPS system, should be readily noticeable. GPS represents the means that has translated the satellite navigation principles into a real system offering a real service: therefore, large emphasis was devoted in Chapter 3 to describing signal structure, radio design, and system architecture. GALILEO represents, instead, an evolution of navigation systems, an ongoing program that pays more attention to user needs (present and future). Therefore, this chapter paints the complex picture of

GALILEO moving from political and programmatic aspects (driven by service requirements), to the architecture for the ground and space segments.

It is worth underlying that all available matter (strategic, political, technical, commercial) related to GALILEO applies to a system that is being designed at the time of writing: the reader has, thus, the advantage of sharing with the authors indeed an advanced topic. On the other hand this matter may change, due to the mandatory (and not fully predictable) needs of system and program optimization; in addition, the references reflect the in-progress nature of the topic and, although public, they are mostly documents. However, in the authors' opinion, this is still the best way to offer to the reader a very special view on the GALILEO world and on the dynamic evolution of its development program.

5.2 The GALILEO Program

The features and planning aspects of the GALILEO program can be understood by being aware of the main guidelines of the European satellite navigation strategy, which has fully recognized how satellite navigation is becoming the primary means of navigation for most civil applications worldwide, thanks to the quality of service offered [10–13]. Satellite navigation, positioning, and timing have already found large application and will be an integral part of the TEN [14]. Many safety-critical service domains (such as multimodal transport) and various commercial applications will depend on this infrastructure. The European Commission White Paper on transport policy highlighted the importance of decoupling economic growth and transport needs, which can be achieved by shifting the balance of transport modes, removing bottlenecks, and placing the user in the core of the transport policy [15]. GALILEO is a promising instrument to reach these goals.

Existing terrestrial radio navigation aids are widespread in number and technology all over Europe. Different types of systems are used by each transport community without a common policy at the European level. In this respect, a European Radio Navigation Plan (ERNP) is being developed to facilitate a common European transport policy. Aviation and maritime areas are already well organized on a global level, while in other cases various national standards are supported. In this context, satellite navigation is a key element of the ERNP, due to its multi-modal and supranational nature.

One of the major concerns for satellite navigation users has been recognized to be reliability and vulnerability of the navigation signal, as the many cases of service disruption reported in the past years—due to unintentional interference, satellite failure, signal denial, or degradation—confirm. In this regard, GALILEO will contribute significantly to reduce those issues by providing additional navigation signals from an independent system, broadcast in different bands. Therefore, recognizing the strategic importance of satellite navigation, its potential applications, and the current shortcomings of GNSS, Europe decided to develop its own GNSS capability in a two-step approach: an augmentation system, EGNOS (see Chapter 4), that is scheduled to be operational in 2005; and GALILEO, an independent navigation system that will be operational starting in 2008 [16].

As previously outlined, GALILEO is a joint initiative of ESA and EC, whereas the name *GalileoSat* has been given to the complementary development program being carried out by ESA; that development includes aspects related to onboard units, constellation control, ground segment data, interference, receivers, TT&C, and test beds. ESA member states adopted the GalileoSat program declaration at Council level (November 2001) and, in the Council of Heads of States and Governments of the EU (March 2002), gave financial support to the GALILEO program. ESA also approved the establishment of the JU to manage the program [16].

The GALILEO program has three dimensions: technological, political, and economic. Satellite radio navigation has become part and parcel of the daily lives of European citizens, featuring not only in their cars and portable telephones but also in their banking habits and the civil protection systems that take care of their security. This picture gives the GALILEO program an additional citizen-friendly dimension.

The total cost of the program is about 3.2 billion euros, with 1.1 billion euros employed in the *development* and validation phase (equally funded by the European Community and ESA) and 2.1 billion euros devoted to the *deployment* phase.

The European Parliament and Council are briefed regularly by the commission on the progress of the GALILEO program. The development phase of GALILEO is at a very advanced stage [17]. 2003 was a decisive year for the program. The JU was set up and commenced its work; the first satellites were ordered; international cooperation was promoted; frequency allocation was confirmed; and the deployment and operating phases were planned. In particular, the JU, whose premises are located in Brussels, has been operational since mid-2003. The staff is composed of some tens of people spread over four departments responsible for technical matters, commercial development, award of concessions, administration, and finance.

The GALILEO concession model for the deployment phase and operation is displayed in Figure 5.1, which outlines the Public-Private Partnership (PPP) concession flow [18].

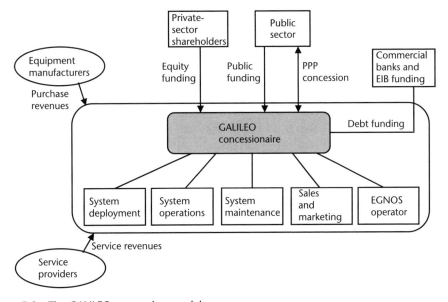

Figure 5.1   The GALILEO concession model.

In June 2003, the European Community achieved important results at the World Radiocommunication Conference. The conference confirmed and set out in detail the conditions governing the use of the frequency spectrum allocated to satellite radio-navigation systems to guarantee the GALILEO operating conditions and, at the same time, protect other major sectors, such as civil aviation [18]. The signal specifications proposed during the definition studies of the system were confirmed. In addition, the position of the European Community in favor of an impartial multilateral coordination by the International Telecommunication Union (ITU) was adopted, hence ensuring, within the allocated frequency spectrum, that distribution among the different systems is not disadvantageous to the European system.

In July 2003, the JU signed an agreement with ESA on the development phase activities related to the ground and space segments. Furthermore, it launched the first call for proposals, under the Sixth Framework Program, and the procedure for awarding concessions for the development and operating phases. Two key studies for the definition phase "Galilei" and "GalileoSat phase B2" were also concluded in 2003 [19]. In particular, "Galilei" has: promoted the presence of GALILEO on the international stage, for example, through organizations such as the ITU and the ICAO; analyzed thoroughly the interoperability with other navigation systems; issued the specifications for the local components of the infrastructure (see Section 5.4); and identified more precisely the international market in equipment and services. "GalileoSat phase B2" has enabled ESA to define in detail the space infrastructure and the associated ground-based segment.

The GALILEO in-orbit validation phase began in July 2003 with the signature, by ESA, of the contracts for the acquisition of the first two experimental satellites; the first is scheduled to be launched by the middle of 2006. By transmitting in orbit prior to the middle of 2006, the two satellites will assure the maintenance of the frequency allocation obtained for GALILEO at the 2000 and 2003 World Radiocommunication Conferences. The two satellites will also contribute to the validation of key onboard equipment, in particular the signal generator and atomic clocks.

Several players involved with satellite radio-navigation have been consulted, such as motor vehicle manufacturers, mobile telephone companies, digital cartography producers, people with reduced mobility, manufacturers of work-site machines, stakeholders in intelligent transport, agricultural and fishing interests, insurers and banking institutions, authorities responsible for civil protection, and the railway community, with the aim of a better understanding the service needs and market potential for the GALILEO system.

GALILEO, due to its unique characteristics, offers new monitoring, control, and management possibilities in a wide variety of sectors, thereby enabling new services and new rules. Therefore, the proposal for a directive on electronic road toll systems makes wide use of the technical capabilities associated with satellite navigation [14]. Electronic road toll systems first appeared in Europe in the early 1990s, with the aim of speeding up toll collection and, hence, capacity of motorways. It was recognized, in fact, that collection of tolls causes congestion, delays, and accidents, which are detrimental to both road users and the environment. Although various systems were introduced, based on short-range microwave technology, at local and then national level, they are mutually incompatible. In view of the growth in international traffic, it is now desirable for these systems to be interoperable at the European level. In this regard, the use of satellite positioning and mobile communications technology is advocated for the deployment of the European electronic toll service, as a key for developing the information society in road transport. It is envisaged that by 2010 the technological progress will render it possible to install in all four-wheeled vehicles equipment communicating with the outside world via microwave, GSM/GPRS, and GNSS interfaces, supporting a range of telematic services, including the electronic toll [14].

International cooperation is an essential element to ensure that maximum benefit is derived from the GALILEO program. Given its range of features, the European satellite navigation system will offer an international public service. Third countries coming forward in ever-increasing numbers asking to be associated with the project have definitively got their priorities right in becoming GALILEO users.

A decisive element in this respect is the agreement that has been reached between the European Union and China on its participation on the GALILEO program [20]. Europe believes that China will help GALILEO become the major world infrastructure for the growing market of location services, and China plans to participate actively in the construction and application of GALILEO for mutual benefit. The CENC (China-Europe global Navigation satellite system technical training and cooperation Centre) was inaugurated in Beijing in 2003, with the aim of becoming a focal point for the activities on GALILEO, as a result of the joint effort by the Chinese Ministry of Science and Technology, the Chinese Remote Sensing Centre, the European Commission, and ESA.

The agreement with China has triggered cooperation requests from other third countries interested in being involved in the development of GALILEO. Cooperation agreements are in progress with India and Israel [21]. Other interested countries include South Korea, Brazil, Japan, Canada, Australia, Mexico, and Chile. Furthermore, Switzerland and Norway, already associated with the GALILEO program through ESA, informed the commission in December 2003 that they were interested in being more closely associated with the program, possibly through involvement in the JU. Further cooperation activities are being undertaken in the Mediterranean region [22], in Latin America, and in Africa.

In July 2003, the commission proposed to the Council and the European Parliament the establishment of two structures for managing the European satellite navigation program: a *Supervisory Authority* and a *Centre for Security and Safety* [23]. The role of the Supervisory Authority is to manage public interests relating to the GALILEO program and, in particular, to act as the licensing authority for the system. It will sign the contract with the Concessionaire and ensure that it is complied with. The Centre for Security and Safety should meet the need for a permanent and decision-making structure whose principal role is to communicate with public authorities and the Concessionaire in the event of a crisis, and even to take measures involving the scrambling of the service signals. It is envisaged that the Centre be placed under the direct responsibility of the General Secretary of the Council/High Representative for the Common Security and Foreign Policy. These two structures will be closely linked to the definition of the security procedures that apply.

5.3   GALILEO Services

GALILEO envisages the provision of a large variety of services [13, 24]. Their definition is based on a comprehensive review of user needs and market analysis. Some of the services will be provided autonomously by GALILEO and others will result from the combined use of GALILEO and other systems. GALILEO services have been classified into four categories:

- GALILEO satellite-only services (GSOS);
- GALILEO locally assisted services (GLAS);
- EGNOS services (ES);
- GALILEO combined services (GCS).

5.3.1   GALILEO Satellite-Only Services

The GSOS services will be provided worldwide and independently of other systems by exploiting the signals broadcast by the GALILEO satellites. All infrastructure elements necessary to provide the five GALILEO services, forming the core of the system, are referred to as the *global component* (see Section 5.4). There is a wide range of possible applications with different operational requirements that have been grouped around the following five service levels:

- *Open service*;
- *Safety of life*;
- *Commercial service*;
- *Public regulated service*;
- Support to *search and rescue* service.

The GALILEO services can be referred to as the latest publicized and accepted realization of the International Terrestrial Reference Frame and the Universal Time Coordinate. This is important for the interoperability with other GNSS, in particular GPS, as will be shown in Section 5.5. The GSOS performance is expressed at the user level; all statistics include the contribution of the receiver (noise, failures, and so on). Users equipped with GALILEO receivers—or having GALILEO functionality in their terminals—conforming to minimum operational requirements shall be able to achieve the specified performance under nominal conditions, without intentional jamming, exceptional interference, exceptional ionosphere and troposphere activity, and with a masking angle of 10° and low multipath environment [24].

A basic service level dedicated to consumer applications and general interest navigation (Figure 5.2) is given by the GALILEO *open service* (OS). The service provides positioning, velocity, and timing information that can be accessed free of direct charge by a small, low-cost receiver. OS is suitable, for example, for in-car navigation and hybridization with mobile telephones. The timing information is synchronized with UTC, when used with receivers in fixed locations, and can be used, for instance, in network synchronization or scientific applications.

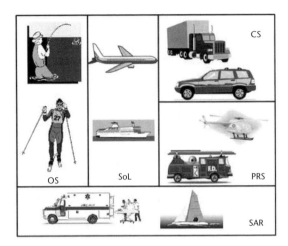

Figure 5.2 Examples of possible GALILEO users.

The OS performance in terms of accuracy and availability will be competitive with existing GNSS systems and further planned evolutions; nonetheless, the service will be interoperable with those systems, in order to ease the provision of combined services. Most receivers will use both GALILEO and GPS signals: this will offer users seamless service performance in urban areas. The OS service performances are listed in Table 5.1 [13, 24]. The OS signals are separated in frequency to allow the correction of errors induced by ionosphere by differentiating the ranging measurements made at each frequency (dual-frequency receiver). The ionosphere correction at the receiver is based on a simple model in the single frequency case. Each navigation frequency will include two ranging code signals (in-phase and quadrature). Data are added to one of the ranging codes, whereas the other "pilot" ranging code is data-less for more precise and robust navigation measurements.

The GALILEO *safety of life* (SoL) service is for use where human safety is critical, as in maritime, aviation, and rail (Figure 5.2). The service will provide high-level performance globally to satisfy the user community needs and to increase safety, especially in areas where services provided by traditional ground infrastructure are not available. As far as transport is concerned, this worldwide seamless service will increase the efficiency of companies operating on a global basis (e.g., airlines, transoceanic maritime companies).

Table 5.1 Positioning and Timing OS Performance

Coverage	Global
Availability	99.8%
Positioning accuracy	H: 15m
(95%, *single frequency*)	V: 35m
Positioning accuracy	H: 4m
(95%, *dual frequency*)	V: 8m
Integrity	No
Timing accuracy (*three frequency*)	30 ns

The SoL service has to deal with the levels of service stipulated by law in various international transportation fields and other recommended practices (e.g., standard and recommended practices, or SARPS, by ICAO). A very specific level of services is needed from GALILEO to comply with legislation applicable to all considered domains of transport and existing standards.

The service will be offered openly and the system will have the capability of authenticating the signal to assure users that the received signal is the actual GALILEO one.

The provision of integrity information at the global level is the main characteristic of this service, if compared with the OS level. The service will be provided globally, according to the performance highlighted in Table 5.2 for the dual frequency receiver [13, 24]. In Table 5.3 the integrity performance requirements are given for each of the three envisaged service levels (A, B, C) [24]. It is worth mentioning that levels B and C are provided on a global coverage, whereas level A is guaranteed on world land masses only, according to the applications highlighted in Table 5.3. The three levels account for conditions of exposure to risk of the transport application of interest. Level A covers time-critical operations, for example, in the aviation domain approach with vertical guidance. Levels B and C are prone to less time-critical applications, such as open sea navigation in the maritime domain.

The SoL signals are separated in frequency to improve robustness to interference and allow correction of errors induced by the ionosphere by differentiating the ranging measurements undertaken at each frequency (dual-frequency receiver). Each navigation frequency includes two ranging code signals (in-phase and quadrature). Data are added to one of the ranging codes, while the other "pilot" ranging code is data-less for more precise and robust navigation measurements. Integrity data will be broadcast in two of the bands assigned to the service (see Section 5.4).

Table 5.2   SoL Service Performance

Coverage	Global
Availability	99.8%
Positioning accuracy (95%, *dual frequency*)	4–6m
Integrity	Yes
Certification/liability	Yes

Table 5.3   SoL Service Integrity Performance

Integrity Level	A	B	C
Integrity availability	99.5%	99.5%	99.5%
Alarm limit (m)	H: 40; V: 20	H: 556	H: 25
Time to alert (TTA) (sec)	6	10	10
Integrity risk (prob/time period)	$3.5 \times 10^{-7}/150$ sec	$10^{-7}/1$ hr	$10^{-5}/3$ hr
Continuity risk (prob/time period)	$8 \times 10^{-6}/15$ sec	TBD /1 hr	$3 \times 10^{-4}/3$ hr
Applications	Aviation APV II, Road, Rail	Aviation en route to NPA	Maritime

The GALILEO *commercial service* (CS) is a restricted-access service level for commercial and professional applications that demands superior performance to generate value-added services (Figure 5.2). The foreseen applications will be based on: dissemination of data with a 500-bps rate, for value-added services; and broadcasting of two signals, separated in frequency from the OS signals to ease advanced applications, such as integration of GALILEO positioning with wireless communications networks (see Chapter 9), high accuracy positioning, and indoor navigation. Further typical value-added applications include: service guarantees; precise timing service; provision of ionosphere delay models; and local differential correction signals for high-precision position determination.

The Galileo Operating Company (GOC) will determine the level of performance it can offer for each commercial service together with verifying the industry demands and the consumer needs. It is intended to provide a guarantee for the service. The CS will be a controlled access service operated by commercial service providers acting after a license agreement between them and the GOC. The CS performance requirements are summarized in Table 5.4 [24].

The CS is based on adding to the open access signals two signals that are protected through commercial encryption. The latter will be managed by the service providers and the future GOC; access will be controlled at the receiver level, using access-protection keys.

GALILEO will provide a further restricted service, the *public regulated service* (PRS), that will be devoted to governmental applications with high continuity characteristics (Figure 5.2). PRS will provide a higher level of protection against the threats to the GALILEO Signal In Space (SIS) than is available for the OS, CS, and SoL service, through the use of appropriate interference mitigation technologies.

The need for PRS results from the analysis of threats to the GALILEO system and the identification of infrastructure applications where disruption to the SIS by economic terrorists, malcontents, subversives, or hostile agencies could result in damaging national security, law enforcement, safety, or economic activity within a significant geographic area. The aim of PRS is to improve the probability of continuous availability of the SIS, in the presence of interfering threats, to users with such a need.

Typical applications include: trans-European level, in particular law enforcement (EUROPOL, customs, European antifraud office OLAF), security services (maritime safety agency), or emergency services (peacekeeping forces or humanitarian interventions); member state level, in particular law enforcement, customs, and intelligence services.

Table 5.4   CS Performance

Coverage	Global*
Availability	99.8%
Positioning accuracy (95%, *dual frequency*)	< 1m
Integrity	*Value-added service*

* Local coverage is also envisaged (see Section 5.3.2)

The introduction of interference mitigation technologies requires that access to these technologies is adequately controlled to prevent their misuse against the interest of member states. Access to the PRS will be controlled through key management systems approved by member state governments. Member states will also control the distribution of receivers.

The service performances are shown in Table 5.5 and the related integrity performances are provided in Table 5.6 [13, 24].

PRS signals are permanently broadcast on separate frequencies with respect to the ones assigned to the open services. They are wideband signals, in order to be resistant to involuntary interference or malicious jamming, therefore offering a better continuity of service.

GALILEO will also provide the European contribution and support to the humanitarian *search and rescue* (SAR) service to accurately pinpoint the location of distress messages from anywhere across the globe (Figure 5.2). The GALILEO SAR service shall: fulfill the requirements and regulations of the IMO, via the detection of emergency position indicating radio beacons (EPIRB) of the Global Maritime Distress Security Service of ICAO via the detection of emergency location terminals (ELT); be backward compatible with the COSPAS-SARSAT (Cosmicheskaya Sistyema Poiska Avariynich Sudov–Search and Rescue Satellite-Aided Tracking) system to efficiently contribute to this international search and rescue effort.

The COSPAS-SARSAT system will take great benefit from the GALILEO SAR service, as it will improve the average waiting time of distress messages transmitted from anywhere, which will become near real time with respect to the current 1-hour figure; improve the location of alert (a few meters for EPIRBs and ELTs equipped with GALILEO receivers with respect to the current 5-km specification on location accuracy); introduce the multiple satellite detection to avoid terrain blockage in severe conditions; increase availability of the space segment [the GALILEO medium Earth orbit constellation adds to four low Earth orbit (LEO) and three geostationary satellites of the current COSPAS-SARSAT system]. In addition, the GALILEO SAR

Table 5.5 PRS Performance

Coverage	Global*
Availability	99–99.9%
Positioning accuracy	H: 6.5m
(95%, *dual frequency*)	V: 12m
Timing accuracy	100 ns
Integrity	Yes

* Local coverage is also envisaged (see Section 5.3.2)

Table 5.6 PRS Integrity Performance

Integrity Availability	99.5%
Alarm Limit (m)	H: 20; V: 35
TTA (sec)	10
Integrity risk (prob/time period)	$3.5 \times 10^{-7}/150$ sec
Continuity risk (prob/time period)	$10^{-3}/15$ sec

service will introduce a new function, namely the return link from the SAR operator to the distress emitting beacon, thereby facilitating the rescue operations and helping to identify and reject the false alarms.

Table 5.7 summarizes the main features of the GALILEO SAR service, including the service *capacity*, intended as maximum number of active beacons each satellite shall relay signals from, and the *forward system latency time*, intended as the interval elapsed from beacon first activation to distress location determination [13]. In fact, the communication from beacons to SAR ground stations shall allow for the detection and location of the distress transmission in a time interval as short as possible in order to be effective.

The SAR GALILEO transponder detects the distress alert from any COSPASS-SARSAT beacon emitting an alert in the 406–406.1-MHz band and broadcasts this information to dedicated ground stations in the dedicated GALILEO frequency window (see Section 5.4). The COSPAS-SARSAT MCC carries out the position determination of the distress alert emitting beacons, once the dedicated ground segment has detected them. The SAR service architecture is displayed in Figure 5.3.

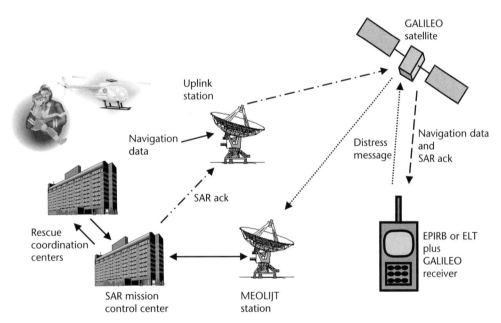

Figure 5.3   GALILEO SAR service architecture.

Table 5.7   GALILEO SAR Service Performance

Availability	99.8%
Capacity (number of active beacons)	150
Forward system latency time	10 minutes
Quality of service [bit error rate (BER) in the beacon-to-SAR ground station link]	$< 10^{-5}$
Acknowledgment data rate (messages per minute)	6 (100 bits long)

5.3.2 GALILEO Locally Assisted Services

Some specific positioning and navigation applications will require very high service performance characteristics that cannot be met by the GALILEO global component alone. Often, these applications are local in nature and require a service within a limited area of coverage. These high-performance requirements will be met by generating additional local signals (*local component*) to augment the satellite ones (similar to the case of 1.5 GNSS systems described in Chapter 4) to provide additional performance in terms of accuracy, availability, continuity, and integrity. The GALILEO local component is part of the overall system definition and consists of GALILEO *local elements*. Both GOC and external service providers will deploy local elements, offering services to a large variety of users. It may also be possible to offer GALILEO local element services guarantees, if the performance characteristics of the local elements of interest meet or improve those of the associated local element standard, once those standards will be defined.

The precise deployment, associated performance, and functionality of local elements will be driven by users and market needs, public regulation, economic factors, and the existing proliferation of networks (e.g., DGPS, GSM, UMTS), which share large infrastructure and functionality required by GALILEO local elements. In this context, four main service categories where local elements will play a role are the following:

- *Local precision navigation services*: GALILEO local elements, providing differential code corrections, will nominally reach position accuracy better than 1m and integrity-associated TTA not higher than 1 second.

- *Local high-precision navigation services*: The exploitation of the TCAR techniques with GALILEO local elements will allow users to determine their position with errors below 10 cm.

- *Local assisted navigation services*: By reducing the amount of decoded information at the receiver, the SIS availability can be improved by, for instance, improved tracking threshold for all GALILEO services, especially when considering applications that operate in difficult environments (e.g., urban canyons and indoor applications). This performance can be further improved by the additional use of the GALILEO OS pilot tones.

- *Local augmented availability services*: Pseudolites (see Chapter 4) will also be used when necessary for increasing the availability of any GALILEO service in a defined local area. In addition, positioning performance will be improved through improved geometry and the fact that the pseudolite signal will not undergo the same level of environmental distortion. Improved availability will be desirable in restricted environments (e.g., urban) and for critical scenarios (e.g., aircraft landing).

The GALILEO local component represents a means to achieve the synergy between the communications and positioning domains, hence capturing the maximum market share.

Almost all GALILEO local elements and associated user terminals will also include additional GNSS and potentially terrestrial-based positioning (e.g., functionality). As a result, the local services offered will be for combined services.

5.3.3 EGNOS Services

According to the integrated strategic vision for the provision of European GNSS, new services can be conceived as a result of combining GSOSs and EGNOS services [25–27]. The latter (see Chapter 4) provides ranging service, wide area differential corrections, and integrity. The combination of GALILEO SoL service with the EGNOS services is of special interest. The combined service will provide independent and complementary integrity information on the GALILEO and GPS constellations that may support, for instance, precision approach type operations in the aviation domain. This ensures that sufficient redundancy exists to offer the potential of sole means availability, avoiding common failure modes between systems, thus allowing the rationalization of the terrestrial traditional radio-navigation infrastructure.

5.3.4 GALILEO Combined Services

GALILEO has been designed to be interoperable with other systems (see Section 5.5) and, therefore, will be widely used as part of a combined service. Combined services result from the combination of GALILEO with either other GNSS systems (GPS, GLONASS, SBAS) or other non-GNSS systems (LORAN-C, GSM, UMTS, INMARSAT, motion sensors, and so on). More details about combination and interoperability are given in Section 5.5.

Combined services bring many benefits: they are able to meet the most demanding user application, reduce satellite navigation system weakness, provide robust solutions for applications requiring system redundancy for safety and/or security reasons, access future GNSS market, and enable and expand new market opportunities.

5.4 The GALILEO System

The design of the GALILEO system has been fully driven by the planned services, described in Section 5.3. A flexible architecture has been preferred to allow for a gradual service implementation. Again, a service-oriented approach has been used to define the components of the GALILEO system. Different portions of the GALILEO infrastructure are needed to provide the services previously described. The GALILEO system components are classified into five groups (according to the services required):

- Global component;
- Local component;
- EGNOS;
- User segment;
- External (GALILEO-related system) components.

The GALILEO system architecture is displayed in Figure 5.4.

5.4.1 Global Component

The global component is the core infrastructure of the GALILEO system that contains all the necessary elements to provide the GSOS. It is composed of the space segment and associated ground segment.

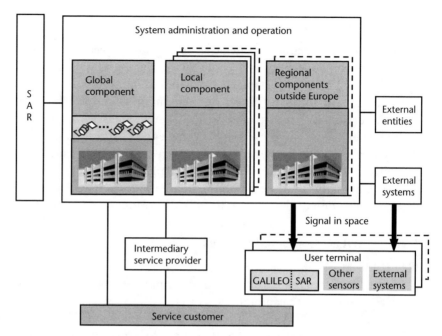

Figure 5.4 GALILEO system architecture.

5.4.1.1 Space Segment

The GALILEO space segment is composed of a constellation of 27 satellites with 3 operational in-orbit spares, in medium height circular orbits (MEO) at 23,616 km above the Earth's surface. Orbits have 56° inclination and a 14-hour, 22-minute period. Satellites are located in three orbital planes equally spaced, forming a Walker 27/3/$\underline{1}$ constellation type (Figure 5.5), where nine operational satellites are equally spaced in each orbital plane and $\underline{1}$ spare satellite per plane is envisaged.

Therefore, orbital and constellation parameters of GALILEO and GPS (see Chapter 3) will be different; at any time and any location on earth the maximum number of visible satellites is 25 (13 from GALILEO and 12 from GPS), 21 (11 from GALILEO and 10 from GPS), and 17 (9 from GALILEO and 8 from GPS) at 5°, 10°, and 15° receiver elevation masking angles, respectively.

The satellites, whose expected design lifetime is 20 years, include a platform and two payloads, one for navigation, the other devoted to search and rescue.

The navigation payload is regenerative and provides the services at L-band. A highly stable timing is achieved through the use of atomic clocks [e.g., Rubidium (Rb), Passive Hydrogen Maser (PHM)]. The navigation payload performs the following functions:

- Reception and storage of the uplinked navigation data;
- Decoding and formatting of integrity data from the integrity link;
- Generation of ranging codes;
- Provision of precise timing information from the onboard clocks;
- Protection of messages with error correcting code;
- Assembly of navigation message in the suitable format;

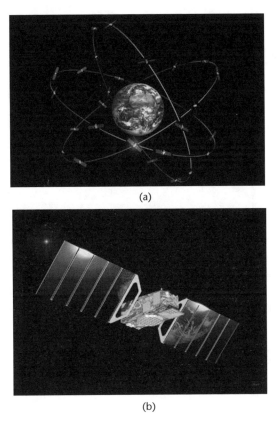

(a)

(b)

Figure 5.5   (a) The GALILEO satellite constellation; and (b) detail of a GALILEO satellite. (© ESA - J. Huart.)

- Broadcast of navigation, timing, and integrity signals, including clock synchronization and orbit ephemeris.

5.4.1.2   Signal In Space

The GALILEO satellite constellation will provide a SIS composed of [28]:

- 10 navigation signals;
- 1 SAR signal.

According to ITU regulations, GALILEO navigation signals will be emitted in the radio-navigation satellite service (RNSS) allocated bands, whereas the SAR signal will be broadcast in the L-band portion reserved for emergency services (1,544–1,545 MHz). The GALILEO navigation signal emissions are summarized in Table 5.8, where the number of signals transmitted in the various frequency ranges and the corresponding labels for the range are provided [24].

The GALILEO SIS description is also graphically displayed in Figure 5.6. Different signals are broadcast on the in-phase (I) and in quadrature (Q) channels and, in the case of the 1,164–1,215 MHz, different signals are provided in the upper (E5b) and lower (E5a) part of the band. Signals are broadcast coherently so that E5a and E5b may be used as a single ultra-wide channel.

Table 5.8   GALILEO Navigation Signal Emissions

Number of Signals	Frequency Range (MHz)	Frequency Range (Label)
4	1,164–1,215	E5a-E5b
3	1,260–1,300	E6
3	1,559–1,591	E2-L1-E1

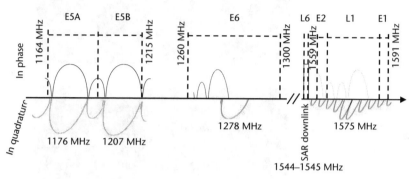

Figure 5.6   GALILEO SIS description.

All GALILEO satellites share the same nominal frequency, making use of CDMA compatible with the GPS approach [29–31]. Similarly to the GPS case (see Chapter 3), each navigation signal consists of a ranging code and data. There are different types of ranging codes and different types of data that can be used for GALILEO signals. The *ranging code* (RC) is a sequence of +1 and –1 with specific characteristics in the time (code length) and the frequency (chip rate) domains. There is one unique sequence for each signal coming from a given satellite. Ranging codes are either publicly known, when the code is actually published, or known only to authorized users, when the code is encrypted.

Three types of ranging codes are envisaged: *open access* RC (publicly known, unencrypted); RC *encrypted* with *commercial encryption*; or RC *encrypted* with *governmental encryption*.

The reference ranging codes consist of a short duration primary code modulated by a long duration secondary code. The resulting code has an equivalent duration equal to the one of the long duration secondary code. Primary codes are based on conventional gold codes with register length up to 25. Secondary codes are given by predefined sequences with code length ranging from 4 (E5b-I channel) up to 100 (E5a-Q and E5b-Q channels). Coded sequence duration ranges from 1 ms (E6b channel) up to 100 ms (E5a-Q, E5b-Q, and E6c channels). An acquisition time in the range of 30–50 seconds is envisaged.

Five types of data are envisaged:

- *Navigation* data;
- *Integrity* data;
- *Commercial* data;
- *PRS* data;
- *SAR* data.

These data are either open access data (navigation data, integrity data, although a capability of integrity data encryption is envisaged, or SAR data) or protected data (commercial data using commercial encryption, PRS data using governmental encryption).

Both RC and data carry the specific information needed for a specific service. Among the 10 GALILEO signals, six are devoted to the OS and the SoL service, two are dedicated specifically to the CS, and two are specifically designed for the PRS.

A half-rate Viterbi convolutional coding scheme is used for all transmitted signals. Given the above frequency plan and target services based on the GALILEO signals, the modulation type of the various GALILEO carriers results from a trade-off among different criteria: minimization of the implementation losses in the satellites, exploiting the current state-of-the-art of the related equipment; maximization of the power efficiency in the satellites; minimization of the level of interference induced by the GALILEO signals in GPS receivers; and optimization of the performance and associated complexity at the user receiver. Modulation of E5 envisages the combination of two quadrature phase shift keying (QPSK) signals. Both E6 and E2-L1-E1 signals contain three channels that are transmitted at the same E6/ E2-L1-E1 carrier, using a modified hexaphase modulation. The latter envisages that a QPSK signal resulting from the combination of two channels is phase modulated with the third channel, and the modulation index (a value of 0.6155 has been chosen) is exploited to set the relative power among the three channels.

The main parameters that have been selected for GALILEO signals are summarized in Table 5.9, together with the channel type mapping into the various services envisaged [24].

5.4.1.3   Ground Segment

The main functions of the ground segment are satellite control and mission control. Satellite control includes constellation management through monitoring and control using the TT&C uplinks. Mission control will globally control the core functions of the navigation mission (orbit determination, clock synchronization) and

Table 5.9   Parameters of GALILEO Signals and Mapping onto Services

Signal ID	Signal Label	Signal Use	Chip Rate (Mchip/s)	RC Encryption	Data Rate (bps)	Data Encryption	Service
1	E5a (I)	Data	10	Open access	25	No	OS/SoL
2	E5a (Q)	Pilot	10	Open access	—	—	OS/SoL
3	E5b (I)	Data	10	Open access	125	Some	OS/SoL/CS
4	E5b (Q)	Pilot	10	Open access	—	—	OS/SoL/CS
5	E6 (A)	Data	5	Governmental	TBD	Yes	PRS
6	E6 (B)	Data	5	Commercial	500	Yes	CS
7	E6 (C)	Pilot	5	Commercial	—	—	CS
8	E2-L1-E1(A)	Data	M	Governmental	TBD	Yes	PRS
9	E2-L1-E1 (B)	Data	2	Open access	125	Some	OS/SoL/CS
10	E2-L1-E1(C)	Pilot	2	Open access	—	—	OS/SoL/CS
11	L6 downlink	Data	—	—	—	—	SAR

determine and disseminate (via the MEO satellites) integrity information (warning alerts within time-to-alarm requirements) on a global basis.

The GALILEO control center (GCC) is the core of the system and includes all control and processing facilities. Its main function includes orbit determination and time synchronization, global satellite integrity determination, maintenance of GALILEO system time, monitoring and control of satellites and services provided, and various off-line maintenance tasks.

Further components of the ground segment are the GALILEO sensor stations (GSS), which collect navigation data from the GALILEO satellites as well as meteorological and other required environmental information. The information is then passed to the GCC for processing.

The GALILEO uplink stations (GUS) include separate two-way TT&C stations in S-band, specific GALILEO mission-related uplinks in the C-band, and the GSSs. The mission uplink stations (MUS) have only mission-related C-band uplinks.

Finally, the global area network (GAN) provides a communication network linking all system elements around the world.

Moreover, a service center is envisaged to provide an interface to users and value-added service providers for programmatic and commercial issues. The center performs the provision of information and warranty on performance and data archiving, information on current and future GALILEO service performance, subscription and access key management, certification and license information, interface with non-European regional components, interface with SAR service providers, and interface with the GALILEO CS providers.

5.4.2 Local Components

The GALILEO local component is composed of all GALILEO local elements. The GALILEO program includes the design and development of some experimental local elements based on specific functionality necessary to meet the associated service requirements. GALILEO local elements will provide, where necessary, enhanced system performance and the possibility to combine GALILEO with other GNSS system- and terrestrial-based positioning and communications systems on a local basis (e.g., D-GNSS, Loran-C, UMTS; see Chapter 9) to a wide variety of users. The following system functionality is required from local element demonstrators:

- *Local precision* navigation elements, providing local differential correction signals that user terminals can use to adjust the effective range of each satellite to correct the ephemeris and clock inaccuracies and to compensate for troposphere and ionosphere delay errors. The quality of the integrity information will be enhanced in terms of both alarm limit and TTA.

- *Local high-precision* navigation elements, providing local differential signals that TCAR user terminals can use to adjust the effective range of each satellite to correct ephemeris and clock inaccuracies and to compensate for troposphere and ionosphere delay errors.

- *Locally assisted* navigation elements, which can use one- or two-way communications functionality to assist the user terminal in position determination in a difficult environment. In a user-terminal-centered approach, one-way communication is required for delivering to the user terminal satellite information (e.g., ephemeris and Doppler) that can be exploited to reduce the time to first fix; this enables the user terminal to determine its own position more quickly from newly acquired satellite signals than would be otherwise possible. This information can also reduce the SIS tracking threshold within the user terminal, which also results in reduced availability. In a service-centered approach, two-way communication is needed to enable received pseudorange information at the user terminal to be transmitted back to a central processing facility, where the position is computed before being retransmitted back to the user terminal.

- *Local augmented-availability* navigation elements, providing local supplementary "pseudolite" transmissions that the user terminal can use as if they were additional GALILEO satellites to compensate for satellite visibility under restricted field of view or high availability requirements scenarios. The local ranging information is also nominally of higher quality than what is received from the GALILEO satellite.

Development of suitable user terminals is envisaged to test and validate the improved performance delivered by each of these local element demonstrators. Implementation, use, and benefit of the local elements will be eased by defining interface control documents (ICD) between the core GALILEO system and external (in particular, mobile communication) systems. The existence of the GALILEO local elements and the proliferation of the mobile communication infrastructure concur to improve the opportunity to conceive applications based on the synergy of two basic functions: navigation and transmission. As a consequence, this synergy will directly translate into the development of the GALILEO market share. Furthermore, service centers might be defined in order to provide to the user community, via local elements, additional value-added services and data (e.g., planned satellite outages, improved ephemeris/clock predictions). The global proliferation of local elements also opens the way to the possible use of the received SIS quality at the local elements to concur with the identification and isolation of interference sources to the GALILEO signals.

It is worth mentioning another "local-based" dimension of the GALILEO system architecture. In fact, the integrity information offered with the SoL is provided by the GALILEO global component and will be valid across the globe. Nonetheless, interested regions can, in addition, provide an optional regional integrity service. This will allow regions to use GALILEO for safety-of-life applications according to their own specific integrity requirements and regulations. Regional components will give regions the possibility to determine their own integrity information and broadcast it via the satellites. The regional-global interface of the GALILEO architecture is displayed in Figure 5.7.

5.4.3 Integration with EGNOS

As highlighted in Chapter 4, EGNOS is composed of four main segments: ground, space, user segments, and support facilities. The EGNOS elements will be kept

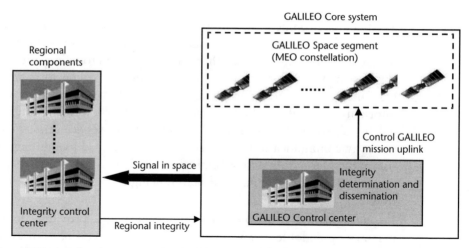

Figure 5.7 Interface between regional and global components of the GALILEO architecture.

functionally independent from the GALILEO global component to avoid common mode of failures.

EGNOS should be used both as a precursor to GALILEO and as an instrument enabling GALILEO to penetrate the market rapidly [26, 27]. The integration of EGNOS into GALILEO does not pose any specific problem: on the institutional level, integrating EGNOS and GALILEO program management into a single entity is the best solution to ensure optimum complementation. Hence, the JU is involved in supervising EGNOS operations and is soon launching the necessary steps to conclude the concession agreement with an economic operator to operate EGNOS starting in 2005.

5.4.4 User Segment

The GALILEO user segment includes the family of different types of user receivers, with different capabilities of using the GALILEO signals according to the various services. Users must be equipped with adequate multifunctional terminals to take full benefit from all GALILEO services (global, local, combined). The functions implemented in the user terminal (UT) should allow the user to: *function 1*) receive directly the GALILEO SIS—this is the actual *GALILEO receiver*; : *function 2*) access services provided by the regional and local components; : *function 3*) be interoperable with other systems. As GALILEO local components, the UTs will be developed in the context of the development phase. The GALILEO receiver, which constitutes the baseline of any GALILEO terminal, performs *function 1*, while *functions* 2 and 3 are optional and depend on the application needs. Some of these functions can be performed by the same physical component (e.g., reception of local components and interoperability with UMTS, interoperability with GPS, and reception of the GALILEO SIS). Some standard terminals will be developed to demonstrate the achievable performance.

The principal scheme and interfaces of the GALILEO UT is shown in Figure 5.8.

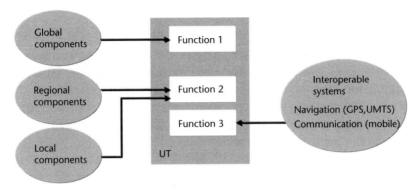

Figure 5.8   Principal scheme and interface of the GALILEO UT.

5.4.5   External Components

As anticipated in Sections 5.1 and 5.2, some non-European countries are undertaking cooperation agreements with Europe in the context of the GALILEO program and further similar initiatives are expected during the operational phase. If non-European regions choose to supplement GALILEO global integrity, the regional components consisting of ground segments, dedicated to GALILEO integrity determination over their specific area, could be envisaged. The deployment, operation, and funding of these components will be under the responsibility of the corresponding regional service providers. The regional integrity data could be routed to the GALILEO ground segment for uplinking to satellites together with the GALILEO and other service provider data.

A further external component of the GALILEO architecture is related to the GALILEO SAR service that, as described in Section 5.3, supports the international COSPAS-SARSAT system. The complete SAR mission consists of: a user segment (distress beacons), which in case of a distress situation transmits an alert message (in the 406–406.1-MHz range); a space segment, which detects the alert messages transmitted by distress beacons and broadcasts them globally in a portion (100 kHz) of the 1,544–1,545-MHz band; a dedicated ground segment (called local user terminals, or LUTs) that receives and processes the alerts related to the space segment. LUTs are designed to receive the alert messages relayed by LEO satellites (LEO-LUTs), geostationary satellites (GEO-LUTs), or MEO satellites like GALILEO (MEO-LUTs); mission control centers, which validate the alert information and distribute it to the rescue team of the rescue coordination centers (RCC).

The contribution of GALILEO SAR service to the international mission consists of the SAR payload onboard the GALILEO satellites and the design of the receiving ground stations (MEO-LUTs). Five stations (distributed adequately) around the world should provide global coverage; the introduction of a new function is envisaged, consisting of a return link from the rescue teams to the distress alert transmitting beacons. The return message is processed by a return link service provider (RLSP). SAR operators (RCC) will designate the RLSP, which will interface with the GALILEO ground segment. The return message will be uplinked by the GALILEO ground segment.

5.5   Interoperability

GALILEO will operate in a scenario where other satellite and terrestrial navigation systems have been or will be operating for many years or decades. As a consequence, interoperability is an issue of fundamental importance for the success of the project.

Interoperability is a functional characteristic of a system, when used in combination with other systems, at the user receiver level, for the provision of new or similar services with enhanced performance [32]. The concept of interoperability is defined into three grades: *coexistence* or compatibility (i.e., absence of interoperability), *alternative use*, and *combined use* (full interoperability).

The *coexistence* of GALILEO with another system means that GALILEO will "do no harm" by degrading the services of another system. This requirement is the main one for any new navigation system and it is assured by the frequency allocations policy of ITU and feasibility study conducted on GALILEO signals [28].

Alternative use of GALILEO with another system means that there is integration at the user receiver level between GALILEO and the other system. The user can use GALILEO or the other system in the same receiver. The user can also use both systems to have new or similar services with enhanced performance.

The *combined* use of GALILEO with another system means that there is a full integration at the system level between GALILEO and the other system. The user can use both systems, as they collapse in a single virtual system, having new or similar services with enhanced performance. This distinction is mainly logical and defines various key grades of interoperability; the actual implementation will show a partial interoperability between systems.

GPS and GLONASS are not interoperable, even if there are GPS and GLONASS integrated receivers on the market. They look like a unique "box" to the user, but the truth is that there are two parallel receivers, which process separated signals and combine them in a manner that potentially still improves user performance. It is important to remark that GALILEO will be fully independent from any other systems, even if partially interoperable with some of them. The GALILEO reference levels for interoperability issues are shown in Figure 5.9 [33].

Finally, interoperability between GALILEO and other systems could be considered into three frames: interoperability with other satellite navigation systems, interoperability with terrestrial navigation systems, and interoperability with non-navigation systems. In this section, the first frame will be addressed and the others will be detailed in Chapter 9. Interoperability between GALILEO and other satellite

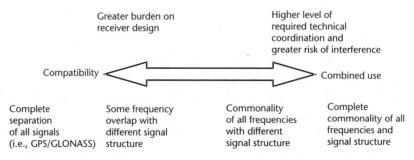

Figure 5.9   GALILEO interoperability levels.

navigation system (i.e., GPS and GLONASS) involves only the first three levels in Figure 5.9.

GALILEO and GPS partially share the allocated frequency, as shown in Figure 5.10. Interoperability will be analyzed with respect to nongovernmental services.

It means that the PRS in GALILEO and the M code in GPS, even if mapped on the same frequency band (L1), will be encrypted and noninteroperable for obvious national and international security reasons. Figure 5.11 shows different signal choices that change the technical burden for the receiver design [34].

Technical issues about interoperability at the signal level involve frequency, bandwidth, spread spectrum and code structure, modulation, data rate, message structure, forward error correction (FEC) algorithm, and power. The GALILEO signal will share the L1 band with GPS and the E5a band will be similar in frequency to GPS L5. Receiver design will benefit from the sharing of a single antenna and it will use an RF channel to compute each for the following signals:

- GPS and GALILEO L1;
- GPS L5 and GALILEO E5b and E5b;

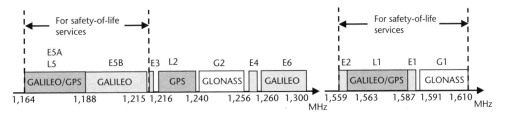

Figure 5.10 Interoperability at spectrum level.

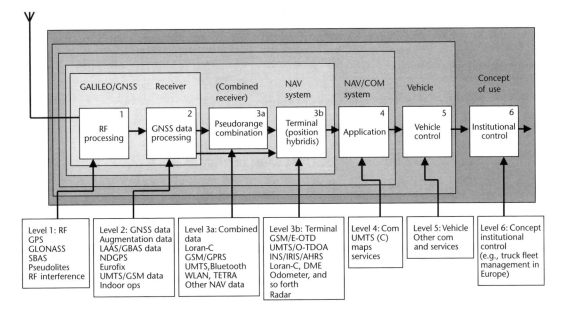

Figure 5.11 GALILEO reference frame for interoperability issues.

- GALILEO E6 (eventually E6 PRS);
- GPS L2;
- Eventually encrypted M code and PRS on L1.

Commercial receivers will compute part of these signals, realistically the first two or three on the list. Cheaper receivers, integrated in mobile phones or for leisure purpose, will compute only the first signal on the list. Receivers for safety-of-life transport should compute the first four signals, whereas military receivers will compute all signals listed.

The digital signal processor of the receiver will process the correlation function between sampled digital signals and locally replicated signals. Therefore, the receiver navigation processor will extract the data that will be computed with the position, velocity, and time (PVT) algorithm. The GPS and GALILEO data must have a common time and geodetic reference system to be merged in a single PVT algorithm. GALILEO will use the GALILEO system time (GST) and it will be steered to the TAI. The offset between GST and TAI shall be kept within 50 ns (2σ) over a one-year period and the offset between GST and TAI shall be known with an uncertainty, less than 28 ns (2σ), over any 24-hour period [35]. GALILEO will broadcast the time difference between UTC and TAI. GPS uses two timing scales, internal "GPS system time" and UTC. The first one is used for position calculation, whereas the second one is used for time calculations for timing users. It is foreseen that GALILEO will broadcast the offset between GST and GPS system time to the users to guarantee the interoperability.

GALILEO will use the GALILEO terrestrial reference frame (GTRF) as geodetic standard. It shall be, in practical terms, an independent realization of the ITRF. GPS uses the WGS-84 coordinate system (see Chapter 2). The differences between the two coordinate systems are within a few centimeters and there is a known mathematical transformation that can be applied to restore the full reference consistency. Most commercial receivers will not need any correction of such a small inconsistency. The navigation processor will receive the pseudorange measurements of GPS and GALILEO, and the PVT algorithm computation can follow two different strategies:

- Hybridization of localization solution;
- Hybridization of raw measurement (pseudorange).

The two methods have strictly the same performance in terms of accuracy [33]. The second method is preferable because the 3D position and time cannot be computed in case the number of visible satellites is less than four and is the case for a single constellation in difficult environments, such as urban canyons. The combined constellations can permit the computation of the 3D position and time if more than three satellites are visible from the two systems.

To properly use GALILEO and GPS in parallel, the receiver shall have sufficient channels to track satellites of both systems.

Political and strategic issues, rather than technical ones, have driven the interoperability decision with GPS that is driving forward [1]:

- Adoption of a common baseline signal structure for their respective open services;
- Confirmation of a suitable baseline signal structure for the PRS;
- A process allowing optimization, either jointly or individually, of the baseline signal structures to further improve performances;
- Confirmation of interoperable time and geodesy standards to facilitate the joint use of GPS and GALILEO;
- Nondiscrimination in trade in satellite navigation goods and services;
- Commitment to preserve national security capabilities;
- Agreement not to restrict use of or access to respective open services by end users.

GALILEO signals are harmonized with the proposed GPS Block III design, which has been based on real operational experience and should be designed to be as interoperable as possible.

The Galilei Studies [33, 36] have shown that the combined processing of GALILEO and GLONASS signals will have a strong impact on the RF receiver. This means that it is not possible with reasonably cheap technologies to process both signals with a single RF channel. It is highly probable that GLONASS signals will not be processed for most applications. Even if the coexistence of the systems will be guaranteed, the GALILEO and GLONASS receivers (if any) are likely to have two parallel receivers (a unique box for users), and the positioning solutions will be computed with separated algorithms.

5.6 Security Aspects

GALILEO was designed for civil users and cannot be considered a dual-use technology. There is no GALILEO document that has so far mentioned a military use of any GALILEO services. For this reason, GALILEO security has not followed the typical procedures of any military system design. Nevertheless, the ownership of a satellite navigation system brings with it added responsibilities. Although the European Commission considers GALILEO a civilian system, other countries or subversive groups or terrorists groups may not. This is why security is so important for GALILEO: it must not be a tool of war and terror.

Because the open service will be unencrypted and free for anyone, the European decision makers must be prepared to deal with the system's eventual unauthorized use by third parties. Nonetheless, several citizen activities and citizen safety will be based on GALILEO. GALILEO must protect itself both from "denial of service" attacks and involuntary interference or malicious jamming that may provide wrong information to the user. The security of GALILEO can be summarized into four key points:

- Security of infrastructure (buildings, stations, space segment, data and communications links between assets). It could result in a reduction of RNP caused by a failure of the assets.

- SIS security (against involuntary interference or malicious jamming). It includes the disruption of PRS or SoL services by terrorists, malcontents, subversives, or hostile agencies.
- Misuse of GALILEO services by nonauthorized users in war scenarios or for terrorist purposes.
- Control of the proliferation of authorized user terminals against smuggling, reverse engineering, and decryption of the transmission keys.

GALILEO is a sensitive infrastructure in terms of security and safety. Since the start of the GALILEO design, security issues have been analyzed by groups of experts of two independent bodies: the GALILEO System Security Board (GSSB) coordinated by the European Commission, and the GALILEO Security Advisory Board (GSAB), set up by the European Space Agency. The GSSB and the GSAB were, in practice, replaced by the GALILEO Security Board (GSB), set up by Article 7 of the Regulation No. 876/2002 to deal with security matters regarding the system. The GSB will have a short existence, as its lifespan coincides with that of the JU and should therefore end with the development phase in 2006. The GSB currently has three main tasks [37]:

- Provide expertise on technical matters regarding security (encryption);
- Assist the commission in its negotiations with third countries by providing expertise (e.g., frequency sharing with the United States);
- Assist with the setup of a future operational framework for security. This includes responsibilities "for the relationship in the event of a crisis to interrupt or restrict signal emission," definition of authorized users, and supervision vis-à-vis international commitments on nonproliferation and export control.

References

[1] "Loyola de Palacio Welcomes the Outcome of EU/US Discussions on GALILEO," IP/04/264, Brussels, Belgium, 25.02.2004.

[2] "Progress in GALILEO-GPS Negotiations," IP/04/173, Brussels, Belgium, June 2, 2004.

[3] "GALILEO: Three Operators Competing for the Concession," IP/04/172, Brussels, Belgium, June 2, 2004.

[4] Ruggieri, M., and G. Galati, "The Space Systems Technical Panel," *IEEE System Magazine*, Vol. 17, No. 9, September 2002, pp. 3–11.

[5] El-Rabbany, A., *Introduction to GPS: The Global Positioning System*, Norwood, MA: Artech House, 2002.

[6] *Progress Report on GALILEO Programme*, Commission of the European Communities, Brussels, Belgium, May 12, 2001.

[7] "Council Resolution of 5 April 2001 on GALILEO," *Official Journal C 157*, 30.05.2001, pp. 1–3.

[8] "Commission Communication to the European Parliament and the Council on GALILEO," Commission of the European Communities, Brussels, Belgium, November 22, 2000.

[9] *Inception Study to Support the Development of a Business Plan for the GALILEO Programme*, Final Report, PriceWaterhouseCoopers, November 14, 2001.

[10] "2420th Council Meeting-Transport and Telecommunications," Brussels, Belgium, March 25–26, 2002.

[11] "Action Programme on the Creation of the Single European Sky," Communication from the Commission to the Council and the European Parliament, Commission of the European Communities, Brussels, Belgium, November 30, 2001.

[12] Iodice, L., G. Ferrara, and T. Di Lallo, "An Outline About the Mediterranean Free Flight Programme," *3rd USA/Europe Air Traffic Management R&D Seminar*, Napoli, Italy, June 2000.

[13] "GALILEO – Mission High Level Definition," EC/ESA, September 2002.

[14] "Developing the Trans-European Transport Network: Innovative Funding Solutions—Interoperability of Electronic Toll Collection Systems"; Proposal for a Directive of the European Parliament and of the Council on "The Widespread Introduction and Interoperability of Electronic Road Toll Systems in the Community," Communication from the Commission, n. COM (2003) 132 Final, 2003/0081 (COD) of April 23, 2003.

[15] "European Transport Policy for 2010: Time to Decide," White Paper, European Communities, 2001.

[16] "Council Regulation (EC) No. 876/2002 of 21 March 2002: Setting Up the GALILEO Joint Undertaking," *Official Journal L 138*, May 28, 2002, pp.1–8.

[17] *Progress Report on the GALILEO Research Programme at the Beginning of 2004*, Communication from the Commission, the European Parliament and the Council, COM (2004) 112 Final, Commission of the European Communities, Brussels, Belgium, February 18, 2004.

[18] "GALILEO Study Phase II – Executive Summary," PriceWaterhouseCoopers, January 17, 2003.

[19] "Inception Study to Support the Development of a Business Plan for the GALILEO Programme," Executive Summary Phase II, PriceWaterhouseCoopers, January 2003.

[20] "EU and China Are Set to Collaborate on GALILEO—The European Global System of Satellite Navigation," IP/03/1266, Brussels, Belgium, September 18, 2003.

[21] "EU and Israel Agreement on GALILEO, Under EU Auspices, Cooperation Between Israel and Palestinian Authority Are Taking Off," DN IP/04/360, March 17, 2004.

[22] "GALILEO Strengthens Euro-Mediterranea Partenrship," IP/03/42, Brussels, Belgium, February 24, 2003.

[23] "Proposal of Council Regulation on the Establishment of Structures for the Management of the European Satellite Radio-Navigation Programme," n. COM (2003) 471 Final, 2003/0177 (CNS) of July 31, 2003.

[24] "The Galilei Project – GALILEO Design Consolidation," European Commission, ESYS Plc, Guildford, United Kingdom, August 2003.

[25] "The Commission Proposes Integrating the EGNOS and GALILEO Programmes," IP/03/417, Brussels, Belgium, March 20, 2003.

[26] "GALILEO – Integration of EGNOS – Council Conclusions," n. 9698/03 (Presse 146), 5.VI.2003.

[27] "Integration of the EGNOS Programme in the GALILEO Programme," Communication from the Commission to the European Parliament and the Council, COM (2003) 123 Final, March 19, 2003.

[28] Hein, G. W., et al., "The GALILEO Frequency Structure and Signal Design," *Proc. ION GPS 2001*, Salt Lake City, UT, September 2001, pp. 1273–1282.

[29] Godet, J., et al., "Assessing the Radio Frequency Compatibility Between GPS and GALILEO," *Proc. ION GPS 2002*, Portland, OR, September 2002.

[30] de Mateo Garcia, J. C., P. Erhard, and J. Godet, "GPS/GALILEO Interference Study," *Proc. ENC-GNSS 2002*, Copenhagen, Denmark, May 2002.

[31] Pany, T., et al., "Code and Carrier Phase Tracking Performance of a Future GALILEO RTK Receiver," *Proc. ENC-GNSS 2002*, Copenhagen, Denmark, May 2002.

[32] Crescimbeni, R., and J. Tjaden, "GALILEO – The Essentials of Interoperability," *Proc. of Satellite Navigation Systems: Policy, Commercial, and Technical Interaction*, Strasbourg, France, May 26–28, 2003.

[33] "GALILEI—Navigation Systems Interoperability Analysis," Gali-THAV-DD080, October 2002.

[34] Turner, D. A., "Compatibility and Interoperability of GPS and GALILEO: A Continuum of Time, Geodesy, and Signal Structure Options for Civil GNSS Services," *Proc. of Satellite Navigation Systems: Policy, Commercial, and Technical Interaction*, Strasbourg, France, May 26–28, 2003.

[35] "GALILEO Mission Requirements Document," Issue 5, Rev. 3, October 10, 2003.

[36] *GALILEI—Multimodal Interoperability Analysis Report*, DD-070, 2002.

[37] Lindstrom, G., and G. Gasparini, "The GALILEO Satellite System and Its Security Implications," Occasional Papers No. 44, Institute of Security Studies—European Union, April 2003.

CHAPTER 6

GPS Modernization Toward GPS III

6.1 Introduction

The GPS was developed by the U.S. Department of Defense to provide precise estimates of PVT to users worldwide [1, 2]. Although the system was initially designed to serve an expected total of 40,000 military users, nowadays the GPS is the mostly widely used navigation system, with more than 20 million civil and military users worldwide.

The growing importance of GPS in both the military and civilian sectors and the need to improve its performance and capability to support high accuracy and high reliability civilian and military services have driven the U.S. government to design a modernization program that will culminate in the implementation of a new navigation system architecture, referred to as GPS III [3–6].

On January 25, 1999, an official release by the U.S. Vice President Al Gore announced a $400 million GPS modernization program [7]. The program's objective is to introduce technological improvements within both the space and control segments and to add new navigation signals to future GPS satellites to improve the performance available for existing services and enable new services in the military, civil, commercial, and scientific sectors worldwide:

> The United States is proud to be a leader in the development of the Global Positioning System—a wonderful example of how technology is benefiting our citizens and people around the world. This initiative represents a major milestone in the evolution of GPS as a global information utility and will help us realize the full benefits of this technology in the next millennium.
>
> —Vice President Gore, January 25, 1999

It is important to highlight that the modernization program will upgrade the GPS by retrofitting the original design without changing the GPS system architecture. However, it is expected that a significant redesign will be required to realize GPS III. Therefore, the modernization process lays down the transition between GPS and the future GPS III, just like GPRS did between GSM and UMTS.

The current GPS modernization program promises to bring to both the military and civil communities numerous improvements to the core of GPS services, which have already enabled so many navigation, positioning, and timing applications in many unexpected ways.

6.2   Space Segment Modernization

As highlighted in Chapter 1, the modernization program started with the conversion of the last eight satellites of Block IIR [Figure 6.1(a)], undertaken by Lockheed-Martin [8], to Block IIR-M, which retrofits Block IIR to include the capability to broadcast the new civil code on L2 frequency, L2C, and also the new military M code on L1 and L2. The first launch of a GPS Block IIR-M spacecraft is scheduled by 2005 and the rest by the end of 2007.

   The current (at the time of this writing) GPS constellation consists of 29 satellites, 2 Block II, 16 Block IIA, and 12 Block IIR. The Block IIR satellites include the last satellite (IIR-12), launched on November 6, 2004, and set usable on November 22, 2004. The IOC for an Earth-coverage M code and L2C code is foreseen in 2014, whereas the FOC is expected to be achieved in 2015.

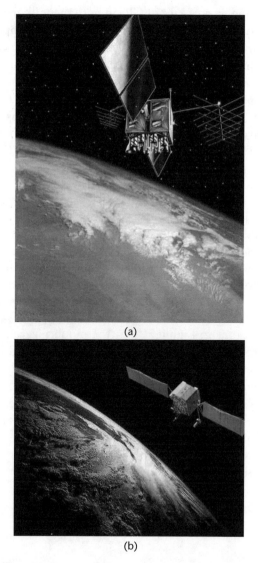

(a)

(b)

Figure 6.1   (a) GPS satellites: Block IIR. (Courtesy of Lockheed-Martin.) (b) GPS satellites: Block IIF.

After Block IIR-M, the "follow-on" satellites will be Block IIF [Figure 6.1(b)], ordered from Boeing North America in 1990 [9–11]. According to the initial agreement, 33 GPS satellites were expected to be built. However, the DoD has reduced the number to 12 in accordance with the current GPS program. The modified Block IIF is referred to as "IIF Lites." Furthermore, the revised agreement provides the introduction of a new navigation payload on the IIF spacecraft, which will enable the capability to broadcast a new third civil signal at 1,176.45 MHz (L5 carrier), as well as the two military M code signals and the civil L2C signal previously introduced into the Block IIR-M. L5 signal will reach IOC in 2016 and FOC in 2019. In addition, it is expected that the Block IIF satellites will have a higher in-orbit life span than the predecessors. The life expectation of a GPS satellite is identified by two parameters: the design life (DL) and the mean mission duration (MMD). Table 6.1 shows the DL and the MMD for GPS satellites of current and future blocks [8, 10–13].

It is worth mentioning that for Block II and IIA the "experienced" MMDs are significantly higher than the "required" ones (see Table 6.1), going up to 9.6 years for the Block II satellites and to 10.2 years for the Block IIA satellites. As a consequence, the current DL and MMD figures for Block IIF are lower than the original ones (15 years for DL and 13 years for MMD) since, considering the increased operational life span of the current satellites, the originally figures would have involved a further delay in the modernization program toward GPSIII [10–13].

Beyond the Block IIR and Block IIF satellites, the current GPS program provides the development of GPS III, whose spacecraft will include all capabilities highlighted for the Block IIF. In addition, it will increase the power levels of the M code signals to improve their antijam proprieties.

The new signal availability is shown in Table 6.2.

The GPS III satellite constellation characteristics are not defined yet. Two ongoing studies are exploring two different architectures: an innovative three-plane constellation and the traditional six-plane constellation used in the current GPS. Also, the number of satellites is still uncertain but should range between 27 and 33.

Table 6.1   Design Life and Mean Mission Durations of GPS Satellites

GPS Block	Design Life (years)	MMD (years)
Block II	7.5	6
Block IIA	7.5	6
Block IIR/IIR-M	10	7.5
Block IIF	12.7	11.3

Table 6.2   New Signal Availability

	L1 C/A	L1 P/Y	L1 M	L2 C	L2 P/Y	L2 M	L5 Civil
Block IIR	Yes	Yes			Yes		
Block IIR-M	Yes	Yes	Yes	Yes	Yes	Yes	
Block IIF	Yes	Yes	Yes	Yes	Yes	Yes	Yes
GPS III	Yes	Yes	Yes	Yes	Yes	Yes	Yes

The nominal schedule for the GPS III high-power code expects IOC and FOC to occur in 2021 and 2023, respectively, with the entire constellation expected to remain operational through at least 2030.

6.3 Control Segment Modernization

The GPS modernization program also involves the control segment. This started in 2000 and is carried out by Boeing with cooperation from Computer Science and Lockheed-Martin. The program's main goals are to reduce the operational costs and operator workload and improve system performance. The modernization process provides an incremental set of upgrades to the OCS that will enhance the system with new functionalities and capabilities. These will take into account the new spacecraft technologies and signals. The main OCS improvements can be summarized as follows [3–6, 14, 15]:

- Addition of accuracy improvement initiative (AII);
- Replacement of the current master control station mainframe computed with a distributed architecture;
- Addition of direct civil code monitoring;
- Building of fully mission-capable alternate master control station (AMCS) at Vandenberg Air Force Base in California;
- Upgraded receiver/antenna and computer technologies within the OCS;
- Addition of IIR-M and IIF control and command capabilities and functionality.

In particular, the AII is an ongoing program started in 1996 by DoD with the aim of leading further improvements in the quality of the broadcast navigation message and then in the GPS overall accuracy. With respect to OCS modernization, the main program initiatives involve an upgrade of the GPS tracking network through the addition of six new NIMA ground stations (Figure 6.2), a new upload strategy, and single-partition processing in the OCS Kalman filter [14].

The tracking network enhancement will result in significant improvement of the timeliness and quality of the GPS tracking measurements as well as related computed parameters. The new upload strategy consists of an increase in the update rate of the navigation data to GPS satellites by OCS.

The nominal GPS modernization schedule for the space and control segment is shown in Figure 6.3, taking into account that the dates indicated for the IOC and FOC could vary depending on the MMD of each satellite block, as experienced with Blocks II and IIA.

6.4 GPS Modernization Signals

The modernization program provides an expanded signal set for civil and military users, ensuring more accurate and reliable measurements. In 1998, Vice President Gore made an official announcement that heralded changes to the L2 signal and

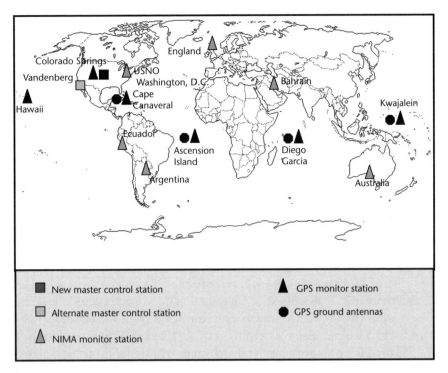

Figure 6.2 Location map of modernized GPS OCS stations.

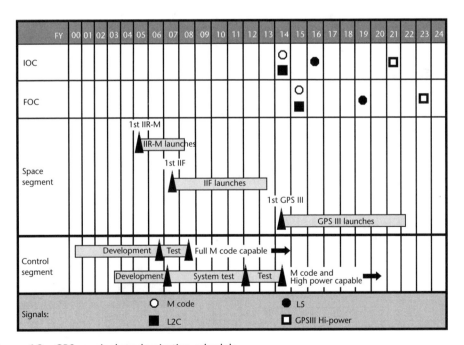

Figure 6.3 GPS nominal modernization schedule.

anticipated the third civil signal for aviation use. The latter was later confirmed by a White House press release [16, 17]. In particular, four new signals will be added, two for civil users on L2 and L5 frequencies and two M code signals for military ones on L1 and L2 carriers. Figure 6.4 shows the evolution of the modernized signals [3–5, 18].

6.4.1 L1 Signal

The modernization program provides the introduction of a new military M code signal at L1 carrier, $f_{L1} = 1,575.42$ MHz.

The L1 signal transmitted by the kth GPS satellite can be expressed in the form:

$$x_{L1,k}(t)\sqrt{2P_{c/a}}C_{C/A,k}(t)D_k(t)\cos(\omega_1 t + \varphi)$$
$$+\sqrt{2P_p}C_{P,k}(t)D_k(t)\sin(\omega_1 t + \varphi) \tag{6.1}$$
$$+ \text{ new military M code}$$

The first two terms in (6.1) refer to the L1 C/A civil signal and L1 P/Y, respectively, previously discussed in Chapter 3. The M code is modulated by a binary offset carrier (BOC) with a subcarrier frequency of 10.23 MHz and spreading code rate of 5.115 Mchip/s, also referred to as BOC(10.23, 5.115) modulation. The spreading code used is a pseudorandom bit stream from a signal protection algorithm. The signal occupies one phase quadrature channel for the carrier, since spreading and data modulations employ biphase modulation.

In addition, the new BOC modulation scheme allows compatibility with existing military and civil signals without producing interference problems, which would reduce the accuracy of the measurements [18–20].

6.4.2 L2 Signal

As previously introduced, new civil and military signals are planned for L2 frequency, $f_{L2} = 1,227.60$ MHz. Considering (3.4), introduced in Chapter 3, the new

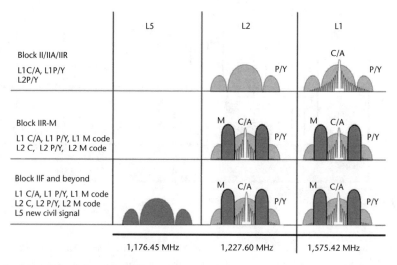

Figure 6.4 Modernized signal evolution.

mathematical expression for the L2 signal transmitted by the kth GPS satellite can be expressed in form:

$$x_{L2,k}(t)\sqrt{2P_p}\,C_{P,k}(t)D_k(t)\cos(\omega_2 t + \varphi)$$
$$+\sqrt{2P_C}\,C_{RC,k}(t)F\{D_k(t)\}\sin(\omega_2 t + \varphi) \qquad (6.2)$$
$$+\text{ new military M code}$$

In (6.2), the first and last terms refer to the military signals. In particular, the first one contains the signal with the P/Y code modulation, previously discussed in Chapter 3, and the last one is the term that introduces the new military signal, modulated by a novel M code.

The second term in (6.2) contains the new civil signal transmitted on L2 carrier and modulated by so-called replacement code (RC). Considering this term, P_C is the signal power, $C_{RC,k}(t)$ is kth satellite spread spectrum RC code, and $D_k(t)$ is the binary sequence of navigation data with a bit rate of 25 bps, whose forward error correction (FEC) scheme is applied with coding rate $R = 0.5$.

The block functional scheme of the L2 civil signal is shown in Figure 6.5 [20].

The new code RC has the same chip rate as the C/A code, 1.023 Mchip/s, and is achieved by time multiplexing two codes, a moderate length code (CM) and a long code (CL) with lengths equal to 10,230 and 767,250 chips, respectively. As result of this, and because both chip lengths are longer than the C/A code chip length, the RC code is significantly longer than the C/A code. As Figure 6.5 shows, the functional scheme provides the possibility for the satellite to broadcast both the RC and C/A codes through a switching control. Moreover, it is important to highlight that the CL code is never modulated by navigation data, as happens with the CM and C/A codes (in this case the possibility to introduce this modulation is regulated by switch). This characteristic is very useful in low signal-to-noise-ratio environments.

The M code signal in (6.2) has the same characteristics as the L1 carrier [18, 19]:

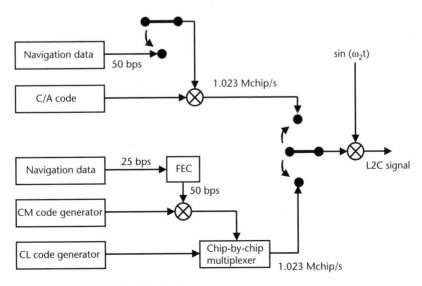

Figure 6.5 L2C signal block functional scheme.

- BOC(10.23, 5.115) modulation with pseudorandom spreading codes;
- Use of biphase modulation in the spreading and data modulations;
- No interference problems with existing civil and military signals.

6.4.3 L5 Signal

Since the current civil signals do not satisfy the needs for some services, including safety critical applications (such as civil air navigation), a new civil signal, called L5, was introduced in the aeronautical radio-navigation services (ARNS) band, f_{L5} = 1,176.45 MHz.

The mathematical expression for L5 signal is expressed here:

$$x_{L5,k}(t)\sqrt{2P_G}\,C_{G_1,k}(t)NH_{10}(t)F\{D_k(t)\}\cos(\omega_5 t + \varphi)$$
$$+\sqrt{2P_G}\,C_{G_2,k}(t)NH_{20}(t)D_k(t)\sin(\omega_5 t + \varphi)$$

(6.3)

Considering (6.3), since no military signal is provided at L5, both in-phase and quadrature components are used for civil purposes. In (6.3), P_G is the signal power, $D_k(t)$ is the navigation data sequence with bit rate of 50 bps, $C_{G_1,k}(t)$ and $C_{G_2,k}(t)$ are the kth satellite spread spectrum codes for the in-phase and quadrature component, respectively, and $NH(t)$ is a term due to the introduction of Neumann-Hoff codes. The L5 functional block scheme is shown in Figure 6.6 [20].

Both spread spectrum codes have a chip rate R_G = 10.23 Mchip/s and, having a L_G = 10,230 code length, their duration is R_G/R_G = 1 ms. Figure 6.6 shows that the in-phase and quadrature components are modulated by Neumann-Hoff codes to extend the code lengths by a factor of 10 and 20, respectively. Considering the chip rate R_G, the new code durations are 10 ms and 20 ms for in-phase and quadrature signals, respectively. Moreover, like the new L2 signal, the L5 carrier also provides a data-free component, since only the in-phase component is modulated by the

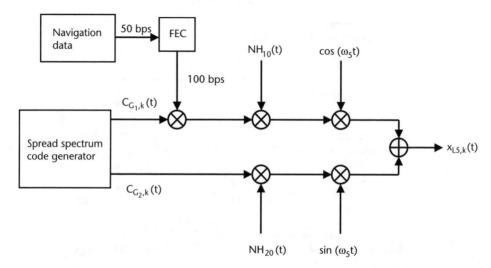

Figure 6.6 L5 signal block functional scheme.

navigation data that, as shown in Figure 6.6, are coded by an FEC scheme with rate $R = 0.5$.

6.5   GPS Modernization Performance and Signal Capabilities

The decision to remove selective availability by the U.S. government can be considered the first step in the modernization process for the civil user environment (see Chapter 3) [21, 22].

In addition, as mentioned in the previous sections, the new civil signals provide innovative properties that result in performance improvements, mainly referred to signal tracking and acquisition [20]. These new properties can be summarized as follows:

- Use of longer codes;
- Introduction of the forward error correction scheme on signals;
- Use of faster code on L5;
- Availability of data-free components.

Each property results in a specific benefit. The use of longer code on L2 and L5 allows one to reduce interference due to correlation and cross-correlation. This feature is very useful in the case of weak signals and, hence, in obstructed environments like cities and forests.

The FEC scheme is applied to the navigation data sequence to reduce the number of errors at the receiver. Using this scheme, performance, evaluated in terms of probability of failed acquisition, improves about 5 dB.

As discussed, the L5 signal has a chip rate value 10 times larger than the L2 signal. The use of faster code improves the characteristics of autocorrelation function, enabling better multipath mitigation and noise reduction within the GPS receivers [23–26].

The introduction of data-free components is particularly helpful in low signal-to-noise ratio (SNR) environments, since it allows users to increase their acquisition capability by using specific algorithms based on prediction schemes.

The availability of more civil signals improves significantly the accuracy performance of the system for users worldwide [27, 28]. Properly combining the civil signals, such as, in the case of three signals availability, through the TCAR method, allows user receivers to reduce ionospheric delay error, which is the main contributor to GPS error budget. In addition, the new L3 civil carrier in the ARNS band provides redundant safety of life services for civil aviation users.

The evolution of the stand-alone GPS UERE budget for the SPS (see Chapter 3) is given in Table 6.3, where the several contributes to final UERE are grouped into user range error (URE) and user equipment error (UEE), according to the related error sources. [15, 29, 30].

According to these considerations and to the OCS modernization initiatives, Table 6.3 highlights that the availability of an additional civil signal will result in a significant decrease of the ionospheric error contribution [27, 28], whereas the availability of additional NIMA stations will reduce the error due to the ephemeris

Table 6.3   Evolution of Stand-Alone GPS UERE Budget for the SPS

Error Source (m)	UERE				
	URE			UEE	
	Atmospheric Delay		Ephemeris and Clock	Receiver Noise	Multipath
	Ionosphere	Troposphere			
L1C/A Availability	7.0	0.2	2.3	0.6	1.5
Two Civil Signals Availability	0.3	0.2	2.3	0.6	1.5
Two Civil Signals and Additional NIMA Stations Availability	0.3	0.2	1.25	0.6	1.5

and clock offsets, allowing then to obtain URE value lower than 1.3m [6, 14]. Therefore, the GPS modernization program will mainly impact the URE contribution. In this respect, Table 6.4 shows the evolution of horizontal stand-alone GPS errors for the SPS at 95% according to the UERE budget outlined in Figure 6.3 and considering a typical value for HDOP.

In light of these considerations, Table 6.5 summarizes the evolution of horizontal accuracy performance of stand-alone GPS for SPS according to the availability of the civil signal [15, 29, 30].

As previously mentioned, the GPS modernization process implies the introduction of M code modulation on L1 and L2 frequencies, which will constitute the core of the military GPS utility for the near future. The main aim of this code is to prevent hostile use of radio-navigation service by unauthorized military users, in order not only to protect the United States and its allies against hostile use of GPS in case of warfare but also to provide continuity of service to peaceful

Table 6.4   Evolution of Horizontal Stand-Alone GPS Errors for the SPS at 95%

Error Source (m)	L1C/A Availability	Two Civil Signals Availability	Two Civil Signals and Additional NIMA Stations Availability
Total URE	7.4	2.3	1.3
Total UEE	1.6	1.6	1.6
Total UERE	7.6	2.8	2.1
HDOP (typical)	1.2	1.2	1.2
Overall horizontal accuracy (m)	18.2	6.7	5.0

Table 6.5   Evolution of Stand-Alone GPS Horizontal Accuracy Performance for SPS at 95%

Time	Horizontal Accuracy (m)	Civil Signal Availability
Prior to May 2000	20 ÷ 100	L1 C/A with SA
Nowadays	10 ÷ 20	L1 C/A with SA off
2015	5 ÷ 10	L1 C/A and L2C
2019	1 ÷ 5	L1 C/A, L2C and L5

users. The code provides innovative cryptography algorithms, more robust signal acquisition, and higher power transmission, which results in an enhancement of performance, evaluated in terms of security, robustness, measurement accuracy, and reliability [31–34].

6.6   GPS III Features

The exponential growth of demand for the radio-navigation service and for increasingly accurate and innovative services foreseen for the next years has led to the necessity of conceiving a new navigation program, which will replace the modernized GPS system. In fact, several studies have demonstrated that the modernized GPS, in spite of the proposed improvements, will not be able to satisfy these demands. The new program, called GPS III, was approved by the U.S. Congress in 2000. The program's initial stages, including requirements definition and architectural design, are currently underway and being run by the Air Force GPS Program Office, together with three contractors: Lockheed Martin, Boeing. and Spectrum Astro.

GPS III, together with GALILEO (see Chapter 5), can be considered the future of worldwide navigation service. The system aims at achieving long-term GPS performance goals and addressing the needs for future civil, military, and commercial needs for the next three decades, providing a cost-effective system able to enhance significantly the quality of the navigation service.

The main GPS III capabilities can be summarized as follows:

- Growth and flexibility;
- High level of signal availability;
- Significant increase in PVT accuracy;
- High level of continuity;
- Significant increase in integrity;
- Autonomous navigation.

The GPS modernization experience demonstrated the difficulty and slowness in obtaining design changes to the operational system. For this reason the necessity to design a system flexible and able to support growing user demand is crucial. This will be obtained by planning a system with the capability to provide product improvements cheaply and rapidly and by using components with a significant margin of flexibility in memory, power, mass, and processor capacity.

GPS III will provide higher power signals. This power increase will improve significantly the system performance, evaluated in terms of measurement accuracy, signal availability, and antijam capabilities. The PVT accuracy performance also depends on the latency figures for the ground segment updates used by the ground control station to transmit ephemeris and clock corrections to the spacecraft. High latency values have the impact of limiting the PVT accuracy. The current GPS provides latency up to 24 hours, updating spacecrafts no more than twice a day. The GPS III design will assure better values of latency, providing updates up to once every 15 minutes. This results in further significant improvement in the stand-alone PVT accuracy.

Moreover, the performance improvements also will be reached by upgrading the ground segment and increasing the robustness and security of ground-to-space and space-to space links.

GPS III will address one of the most important weaknesses of the current stand-alone GPS system (i.e., integrity), which is a paramount requirement for safety critical applications such as civil air navigation. In this respect, GPS III will ensure navigation solution integrity by performing outage monitoring, detection, validation, alerting, and the initiation of corrective action. This will be obtained through a worldwide network that will continuously monitor the state of SIS, providing a timely alert to users in case of unacceptable degradation of the signal quality.

Ongoing studies, based on the Iridium, TDRS, and Silex experiences conducted by The Aerospace Corporation, leads one to consider a space-based network using intersatellite cross-links rather than widening the ground station network to satisfy the requirements mentioned [35, 36].

Autonomous navigation means the satellite capability to autonomously estimate clock and ephemeris corrections. The introduction of intersatellite link in the space segment will allow users to meet this requirement of self-monitoring by using onboard cross-link-ranging measurements. This function will considerably reduce workload of the master control station and, in case of need (e.g., control station outage), can assure the maintenance of high navigation service performance.

The implementation of these capabilities requires a network design with a capability to detect and report sudden failures or outages within a specified time period, to broadcast data from sensors, and to allow fast constellation control and relay the navigation data message.

The GPS III architecture is shown in Figure 6.7. In summary, GPS III will improve overall performance, increasing war fighter capability and, at the same

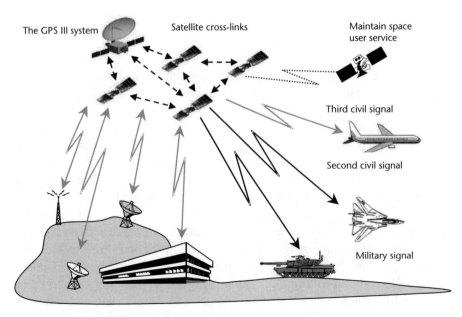

Figure 6.7 GPS III system architecture.

time, satisfying future civil needs, thereby ensuring the best GPS performance through 2030.

References

[1]   Kaplan, E., (ed.), *Understanding GPS: Principles and Applications*, Norwood, MA: Artech House, 1996.

[2]   Parkinson, B. W., and J. J. Spilker Jr., "Global Positioning System: Theory & Applications," *American Institute of Aeronautics and Astronautics*, Vol. I, 1st edition, January 15, 1996.

[3]   Fontana, R., and D. Latterman, "GPS Modernization and the Future," *Proc. of IAIN/ION Annual Meeting*, San Diego, CA, June 2000.

[4]   Van Dierendonck, A. J., and J. Spilker, "GPS Modernization," *Journal of the Institute of Navigation*, 2001.

[5]   Shaw, M., K. Sandhoo, and D. Turner, "Modernization of the Global Positioning System," *GPS World*, Vol. 11, No. 9, September 2000, pp. 36–44.

[6]   Evans A. G., and R. W. Hill, (eds.), G. Blewitt, et al., "The Global Positioning System Geodesy Odyssey," *Journal of the Institute of Navigation*, Vol. 49, No. 1, Spring 2002, pp. 7–34.

[7]   "New Global Positioning System Modernization Initiative," Office of the Vice President, White House press release, January 1999.

[8]   http://www.lockheedmartin.com.

[9]   "The Boeing Company Receives GPSIIF Modernization Approval,"Announcement—The Boeing Company: Key Developments, March 13, 2002.

[10] Fisher, S. C., and K. Ghassemi, "GPS IIF-The Next Generation," *Proc. of the IEEE*, Vol. 87, pp. 24–47, No. 1, January 1999.

[11] http://www.boeing.com/satellite.

[12] http://www.navcen.uscg.gov.

[13] de Jong, K., "Success Rates for Integrated GPS and GALILEO Ambiguity Resolution," *Revista Brasileira de Cartografia*, No. 54, Brazil, December 2002.

[14] Malys, S., et al., "The GPS Accuracy Improvement Initiative," *Proc. of ION GPS-97*, Kansas City, MO, September 1997.

[15] Galati, G., "Detection and Navigation Systems/Sistemi di Rilevamento e Navigazione," Italy: Texmat, 2002.

[16] "GPS to Provide Two New Civilian Signals," Office of the Assistant Secretary for Public Affairs, Press Release, March 1998.

[17] Spilker, J. J., and A. J. Van Dierendonck, "Proposed New Civil GPS Signal at 1176.45 MHz," *Proc. of ION GPS-99*, Institute of Navigation, Nashville, TN, September 1999.

[18] Betz, J. W., et al., "Overview of the GPS M Code Signal," *Navigating into the New Millennium: Institute of Navigation Nat. Tech. Mtg.*, Anaheim, CA, January 2000, pp. 542–549.

[19] Barker, B. C., et al., "Details of the GPS M Code Signal," *Proc. of ION 2000 National Technical Meeting*, Institute of Navigation, Long Beach, CA, January 2000.

[20] Enge, P., "GPS Modernization: Capabilities of the New Civil Signals," Australian International Aerospace Congress, Brisbane, Australia, August 2003.

[21] Georgiadou, Y., and K. D. Doucet, "The Issue of Selective Availability," *GPS World*, Vol. 1, No. 5, September/October, 1990, pp. 53–56.

[22] Conley, R., "Life After Selective Availability," *U.S. Institute of Navigation Newsletter*, Vol. 10, No. 1, Spring 2000, pp. 3–4.

[23] Bétaille, D., et al., "A New Approach to GPS Phase Multipath Mitigation," *Proc. of ION National Technical Meeting*, Anaheim CA, January 2003, pp. 243–253.

[24] Kelly, J. M., and M. S. Braasch, "Validation of Theoretical GPS Multipath Bias Characteristics," *Proc. of IEEE Aerospace Conference*, Big Sky, MT, March 2001.

[25] Park, K., et al., "Multipath Characteristics of GPS Signals as Determined from the Antenna and Multipath Calibration System (AMCS)," *Proc. of ION GPS Meeting*, Portland, OR, September 2002, pp. 2103–2110.

[26] Ray, J. K., M. E. Cannon, and P. Fenton, "GPS Code and Carrier Multipath Mitigation Using a Multiantenna System," *IEEE Trans. on Aerospace and Electronic Systems*, Vol. 37, No. 1, January 2001, pp. 183–195.

[27] Afraimovich, E. L., V. V. Chernukhov, and V. V. Dernyanov, "Updating the Ionospheric Delay Model Using GPS Data," Application of the Conversion Research Results for International Cooperation, *Third International Symposium, SIBCONVERS '99*, Vol. 2, Tomsk, Russia, May 1999, pp. 385–387.

[28] Hatch, R., et al., "Civilian GPS: The Benefits of Three Frequencies," *GPS Solutions*, Vol. 3, No. 4, 2000, pp. 1–9.

[29] Kovach, K., "New User Equivalent Range Error (UERE) Budget for the Modernized Navstar Global Positioning System (GPS)," *Proc. of The Institute of Navigation National Technical Meeting*, Anaheim, CA, January 2000.

[30] Lau, L., and E. Mok, "Improvement of GPS Relative Positioning Accuracy by Using SNR," *Journal of Surveying Engineering*, Vol. 125, No. 4, November 1999, pp. 185–202.

[31] Betz, and J. W., "Analysis of M Code Interference with C/A Code Receivers," *Proc. of ION 2000 National Technical Meeting*, Institute of Navigation, Long Beach, CA, January 2000.

[32] Betz, J. W., "Effect of Jamming on GPS M Code Signal SNIR and Code Tracking Accuracy," *Proc. of ION 2000 National Technical Meeting*, Institute of Navigation, Long Beach, CA, January 2000.

[33] Betz, J. W., and J. T. Correia, "Initial Results in Design and Performance of Receivers for the M Code Signal," *Proc. of ION 2000 National Technical Meeting*, Long Beach, CA, Institute of Navigation, January 2000.

[34] Fishman, P., and J. W. Betz, "Predicting Performance of Direct Acquisition for the M Code Signal," *Proc. of ION 2000 National Technical Meeting*, Institute of Navigation, January 2000.

[35] Maine, K., P. Anderson, and F. Bayuk, "Communication Architecture for GPS III," *IEEE Aerospace Conference*, Big Sky, MT, March 2004.

[36] Maine, K., P. Anderson, and F. Bayuk, "Cross-Links for the Next-Generation GPS," *IEEE Aerospace Conference*, Big Sky, MT, March 2003.

CHAPTER 7

Legal and Market Policy of Satellite Navigation

7.1 Introduction

An important step in understanding the features and perspectives of the satellite navigation world can be achieved by an awareness of the various system interfaces and issues occurring at legal and market-related levels.

Forecast of future global markets for navigation show that this industry is at the beginning of a large expansion, with an envisaged global turnover of about 140 million euros by 2015 [1]. Initially characterized by a strong product-oriented dominance, it is expected that service provision will rapidly play an important role in the satellite navigation market.

Figure 7.1 shows the annual turnover directly associated with positioning systems hardware (net turnover) in the 2000–2020 time scale [1].

The trends of the global annual satellite navigation product and service turnovers are displayed in Figure 7.2. It is envisaged that service turnover will grow to 112 billion euros by 2020 [1].

Regulation at a number of levels—international, continentwide, and national—will steer the use of satellite navigation systems.

In general, regulation either mandates performance or technology or authorizes the use of a certain technology. *Mandate of performance* regulations require provision of service with a set of performance criteria and are neutral from a

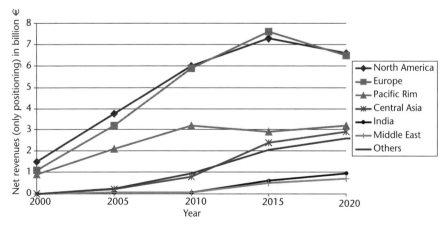

Figure 7.1 Annual net turnover for satellite navigation products.

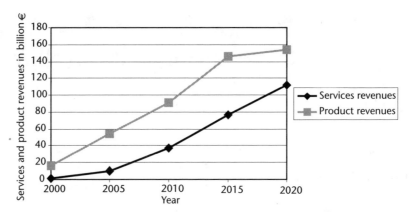

Figure 7.2 Global annual satellite navigation product and services turnover.

technological point of view. A typical example is the U.S. regulation mandating localization of calls to emergency services from mobile phones; to comply with the related Federal Communications Commission (FCC) regulations. Several U.S. carriers have adopted GNSS-based solutions. Whereas AT&T and Cingular Wireless chose network-based solutions, Cingular, Nextel, Sprint PCS, and Verizon Wireless have opted for assisted-GPS. The latter companies are scheduled to provide 95% of all subscriber handsets in service nationwide with GPS capability by December 31, 2005 [1]. Considering that Nextel, Sprint PCS, and Verizon Wireless have more than 50 million wireless users overall, the quoted regulation will improve considerably the market uptake of satellite navigation.

In the case of *mandate of technologies*, regulations require the use of a particular technology for provision of services. This obviously boosts the mandated technologies. The existence of such regulations gives GNSS a certain and well-defined market boost over other technologies. An example is given by the British government policy to equip all ambulances with satellite navigation units for resource management [1]. A further example is the German toll system for truck journeys on motorways, collected through satellite navigation-based systems. It is estimated that 1.2 to 1.4 tracks will be subject to this toll and, hence, they will be fitted with satellite navigation receivers [1]. Concerning *technology authorization*, there are cases where the use of satellite navigation is not mandatory but is recommended as a standard navigation aid. This is, for instance, the case of the SARPs in the aviation Chicago Convention. Many countries, in particular South America and Africa, normally transpose these SARPs into national legislation without major modifications, leading to direct positive impact on receiver sales. International bodies like ICAO and IMO are elaborating future policies based on the use of satellite navigation. In Europe, regulation is either being implemented or under discussion in various domains, such as road tolls, agriculture, fisheries, road (e-safety), customs, justice and home affairs, environment, and telecommunications. The new policies and standards will drive the demand for accurate and reliable navigation systems [2].

Given these complex situations, the aim of this chapter is to provide the reader with an overview of the legal and market-related issues, focusing on the GPS, augmentation, and GALILEO systems. The legal and market-related policies governing these systems are influenced by different development and operational stages. As a consequence, the authors have chosen to highlight those aspects that provide an effective picture of the specific system, related achievements, and trends.

In particular, as GPS represents the practical implementation of the navigation concepts, the original approach to the positioning and navigation market and the progress in the analysis are particularly interesting, as well as the aspects related to standardization.

The augmentation systems in turn have legal and market aspects that derive from their intrinsic nature of bridging two generations of navigation systems. Finally, GALILEO is experiencing the legal and market issues related to its status of "system under development," compressed by time and financial constraints.

All these aspects have been taken into account in the selection and development of the topics included in this chapter.

7.2 GPS

The GPS system structure was described in Chapter 3. This section aims at giving readers additional information about the system, referring, in particular, to the U.S. legal and market policy of radio-navigation, since these represent key aspects for the system's maintenance and further development.

7.2.1 Market Aspects

GPS was originally deployed to aid U.S. armed forces in navigation and positioning. However, over the years, the system has evolved far beyond its military origins, providing navigation services to both military and civilian users. GPS can now be considered an information resource supporting a wide range of civil, scientific, and commercial functions with precision location, timing, and velocity information. Considering the large number of current GPS civil applications, the market for civilian users nowadays exceeds its military counterpart by a ratio of approximately 8 to 1 [3–5].

The success of GPS and the growth in civilian applications has produced a booming market for GPS services and products. Figure 7.3 shows a forecast of GPS equipment revenue between 2000 and 2008 (inclusive), as highlighted in an Allied Business Intelligence (ABI) study [6].

This significant growth is essentially due to two factors: the consistent reduction of the cost of GPS receivers, due to new smaller and cheaper technologies, and the increase in the services provided to users. Figure 7.4 shows the trend in the cost of GPS receivers in the period 1991 to 2004.

Therefore, moving from a system conceived for military purposes without any commercial ambition, the GPS now moves a volume of business worth billions of dollars. This volume is expected to grow in the next decades, considering, in particular, the GPS modernization program and the new capabilities that will be added to the system (see Chapter 6).

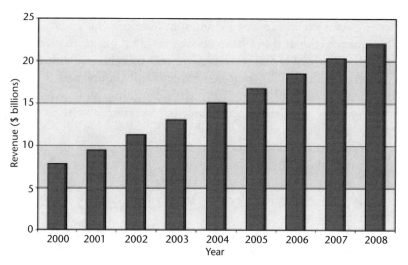

Figure 7.3   Forecast of GPS equipment revenue.

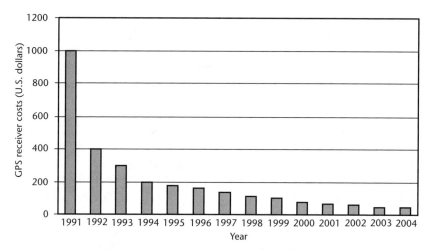

Figure 7.4   GPS receiver cost trend.

7.2.2 Cost Aspects

GPS is an expensive system referring, in particular, to the space and control segments, along with their operation, replacement, and maintenance. For this reason, the U.S. government is obliged to perform continuously an accurate analysis of the system costs, evaluating the system budget and avoiding unnecessary costs.

Because GPS currently provides commercial services, it can be considered a commercial system and, hence, as the market laws demand, it has to ensure profits. The continuous cost analysis is the basis for evaluating and setting priorities for investment in new services. Hence, the U.S. economic policy aims at maintaining in operation only cost-effective systems, namely, systems that are able to provide economic benefit.

GPS supports applications related to safety, security, and commerce. The economic benefit of these navigation services is evaluated on a national scale, affecting, hence, the national economy. In some instances, such as air navigation in areas where federal systems are not justified, provision of aids to navigation is not economically profitable for the U.S. federal government. In these cases, navigation services are supported by private, corporate, or state organizations that yield economic benefit to the local economy.

U.S. federal government policy provides service cost recovery through either general tax revenues or transportation trust funds, which are generally financed via indirect user fees. However, as stipulated by presidential decision directive, the GPS SPS (see Chapter 3) has to be provided free of charges to all users.

7.2.3 GPS Standard Formats

This section considers standards as applied to the handling and management of GPS data. This is important because standards constitute a boundary area in system development and operations, due to their impact on both market and legal aspects of system health and effectiveness.

The considerable growth of the GPS market has led to the need for both manufacturers and users to define common standard formats that ensure full interoperability, compatibility, and integrability among the different equipment and devices. These standard formats have been developed by a number of research groups to meet the already mentioned needs. A brief description of each of the most commonly used formats is given here.

7.2.3.1 RINEX Format

The Receiver IndepeNdent Exchange (RINEX) is the international standard format, developed by a group of researchers at the University of Berne I Switzerland, to exchange between different GPS receivers [7]. RINEX is an ASCII file and this characteristic increases its distribution flexibility, since ASCII files are simple text files easy to read for any commercial GPS receivers (and even for people). A RINEX file, in particular, corresponds to a translation of the compressed binary files stored in the receiver memory. Different version of RINEX format has been developed; the current one is version 2.10, which defines the following six RINEX files:

- Observation data file;
- Navigation message file;
- Meteorological file;
- GLONASS navigation message file;
- Geostationary satellites data file;
- Satellite and receiver clock data file.

A new version 2.20 is currently proposed to deal with LEO satellites equipped with GPS/GLONASS receivers [8]. The naming convention for this standard format is "ssssdddf.yyt," whose meaning is explained in Table 7.1.

Considering these RINEX file types, the most important for the majority of GPS users are the observation data, the navigation message, and the meteorological files. Each of these includes two sections: a header and data. In particular, the

Table 7.1   Explanation of the RINEX Naming Convention

Characters	Meaning
ssss	Site designation
ddd	Day of year of the first record
f	File sequence number
yy	Two-digit year
t	File type: O – observation
	N – navigation
	M – meteorological
	G – GLONASSH – geostationary

observation data file contains information on measurement (observation) session, receiver/antenna station, GPS time, measurements (raw code and carrier data), number, and list of satellites in view during the session. The navigation message file includes the satellite information contained in the navigation data highlighted in Chapter 3, such as satellite clock parameters, satellite health, and almanac parameters. The meteorological file contains time-tagged information on atmospheric status, such as temperature, pressure, humidity, and other related information.

7.2.3.2   NGS-SP3 Format

As described in Chapter 2, a key aspect for evaluating the user position is the precise knowledge of the satellite ephemeris. To facilitate exchanging precise satellite ephemeris, the National Geodetic Survey (NGS) developed the NGS-SP3 format [9, 10]. The Standard Product 3 (SP3), in fact, is the international standard that regulates the format of satellite information such as precise orbital data and the associated satellite clock corrections, taking into account that all times are referred to GPS system time. The NGS-SP3 is a 60-character-long ASCII file organized in two sections: header and data. The header includes information on the observation session such as date and number of satellites. The data are organized in epochs (records). Each epoch contains ephemeris and clock corrections of any satellite of the GPS constellation, assigning a specific line of the ASCII data file per each satellite.

7.2.3.3   RTCM SC-104 Format

The Radio Technical Commission for Maritime Services, Special Committee 104 (RTCM SC-104) format is the industry standard related to DGPS services [11]. The format was proposed by RTCM to ensure and improve the efficiency of the differential operations. This standard, in fact, defines the format of differential pseudorange corrections transmitted by the reference stations to the rover receivers (see Chapter 4). Hence, it is used to transmit information for real-time DGPS. The RTCM SC-104 includes 64 message types, each including differential information. The messages consist of streams of binary digits whose length changes depending on the message content and type. Among the 64 message types, the most interesting messages for real-time DGPS users are types 1 and 9, since both include two very useful differential information for the user pseudorange measurements, namely, pseudorange correction (PRC) of each satellite in view of the transmitting reference station and the rate of change of pseudorange corrections (RRC).

7.2.3.4 NMEA 0183 Format

The National Marine Electronics Association (NMEA) 0183 format is the standard used in real-time marine navigation to interface marine electronics devices. Particularly, this standard format is used for transmitting GPS information from the GPS receiver to hardware that uses the positioning as input [12]. The NMEA 0183 standards are data streams in the ASCII format and the data are transmitted in the form of sentences, each having no more than 82 characters.

7.2.4 GPS Federal Authorities

The GPS is regulated by the U.S. federal government and, taking into account the "dual use"—civilian and military—of the system, two government authorities manage the system and the services provided: the Department of Defense (DoD) and the Department of Transportation (DOT) [3, 13].

The DoD has the responsibility to operate and maintain aids to navigation required solely for national security and defense. In particular, the DoD develops, tests, evaluates, implements, operates, and maintains aids to navigation and user equipment required exclusively for national defense. In addition, the DoD ensures the necessary navigation capabilities required for military vehicles, ensuring interoperability between the military and civilian environments. The DoD also has the responsibility of providing free GPS SPS to civil users worldwide, as established by statute 10 U.S.C. 22881 [3, 13].

The role of the DOT is to provide [under title 49 United States Code (U.S.C.) Section 301] navigation aids to ensure efficient and safe transportation. Within DOT three main federal agencies participate in radio-navigation planning:

- The Federal Aviation Administration;
- The U.S. Coast Guard (USCG);
- The St. Lawrence Seaway Development Corporation (SLSDC).

The FAA is responsible for developing and implementing navigation systems to enable efficient and safe air navigation in the National Airspace System, both for military and civil aviation. The FAA also provides aids to air radio-navigation required by international treaties. The USCG is responsible for operating navigation aids for efficient and safe marine navigation. The SLSDC operates aids to navigation in U.S. and St. Lawrence River waters. The SLSDC also provides a vessel traffic control system in cooperation with the St. Lawrence Seaway Authority of Canada. The DOT includes several other federal authorities [e.g., the Intelligent Transportation Systems Joint Program Office (ITS-JPO), the Maritime Administration (MARAD), and the Federal Transit Administration (FTA)] that provide navigation aids to specific users.

In addition, other federal agencies participate in radio-navigation planning, such as NASA, the NGS, the National Oceanic and Atmospheric Administration (NOAA), and the Department of Commerce (DOC).

7.3 Augmentation Systems

This section presents an overview of the legal and market policies relevant to the WAAS and EGNOS satellite-based augmentation systems (see Chapter 4). Concerning EGNOS, special attention is paid to ongoing discussions on its integration with GALILEO. This is important because of its relevance to the understanding of evolution of the European radio-navigation plan.

7.3.1 WAAS

As highlighted in Chapter 4, the WAAS is a GPS-based navigation and landing system designed primarily for air navigation users. The WAAS signal in space is currently available, providing a reliable navigation service for most of the United States. Although the WAAS is still under development, its benefits are already available on the market and, hence, to users.

To better understand the impact of WAAS on the airborne market, it is worth mentioning some elements about this market. The air navigation market can be basically divided into two categories: the air carrier industry and the general aviation community. The latter refers to private aircraft owned by individuals or companies for personal or recreational flying, as well as corporate transportation. Those markets move a business volume of billions of dollars and are characterized by a high demand for radio-navigation services [4]. The WAAS aims at satisfying this demand by providing aids to air navigation with improved accuracy, availability, and integrity of GPS. As a consequence, the WAAS could replace most of the navigation systems currently employed in the air navigation environment, containing their navigation capabilities into a single system. This would reduce controller workload and, especially, avionics costs, leading to more consistent and efficient operations [14]. In particular, referring to the avionic cost issues, the WAAS capabilities substantially reduce the aircraft and training costs, since the WAAS reduces the number of navigation instruments required onboard to ensure the high standards provided by the instrumental flight rules (IFR) and the safety and capacity provided by the NAS.

In this respect, the WAAS represents a good investment opportunity by both the federal government and private sector. Nonetheless, in September 2003, the U.S. Senate Appropriations Subcommittee replaced the House Appropriations Subcommittee recommendation to fully fund the WAAS program and considerably reduced and restricted the WAAS funding [15–17]. It is easy to forecast that this budget cut will result in a significant slowdown of the WAAS development, and, hence, the full availability of WAAS benefits for both public and private users will be delayed.

The current WAAS market policy is oriented to an international expansion, since the FAA currently authorizes the use of the system only within NAS. In fact, due to this geostationary satellite-based structure, the WAAS cover includes large portions of Canada and Mexico. Therefore, with the addition of few ground reference stations, the WAAS capabilities could be provided in those countries. The WAAS team is working to support this effort, since the involvement of Mexico and, especially, Canada in the WAAS program represents an important business opportunity for the U.S. federal government [18].

7.3.2 EGNOS

EGNOS, the European augmentation system (see Chapter 4), is a key building block of the evolutionary scenario of navigation services suggested by GALILEO (see Chapter 5). Estimated benefit from EGNOS on a stand-alone basis has been quantified for five applications based on the potential for further business development, estimating a total accrued net benefit—mostly expected in Europe—up to 2020 for nonaviation applications of about 15 billion euros [19]. The benefit share (in percentage) of the identified five applications is summarized in Table 7.2.

In addition to the highlighted five sectors, many potential benefits of EGNOS are expected in the civil aviation community. The commercial airline community is only part of the civil aviation sector where revenues might be derived: charges could be based on existing Eurocontrol mechanisms, this being consistent with estimated cost-savings from decommissioning VHF omni-ranging and nondirectional beacons. Forecasts of revenues from the stand-alone EGNOS business are not envisaged to rise above operating costs in the next 16 years: hence, the program would need further financial support to be viable. Any private company expecting a financial return is, thus, only going to accept the EGNOS contract with guaranteed financial support from the public sector, that may, hence, wish to integrate EGNOS and GALILEO to gain the benefits of cost reductions that derive from integration.

Although the EGNOS development has been driven mainly by civil aviation requirements, with services to a core region equivalent to the union of the ECAC and the European Maritime Core Area, there are further strategies for additional EGNOS business development: geographical expansion, market sector expansion, and exploitation of delivery mechanisms other than geostationary satellites (see Chapter 4). These alternative mechanisms include the Internet, digital audio broadcast, and mobile telephony, all effective in urban environments where geostationary satellite visibility can be limited.

An important aspect of EGNOS market analysis concerns its impact on the GALILEO system [19–22]. The benefits of EGNOS to GALILEO have been

Table 7.2   Benefits From EGNOS

Market Sector	Benefit Share (%)
Road transport	84
Precision farming	12.6
Hydrographical survey	2.0
Train protection and control	0.7
Inland waterways	0.7

identified in four major categories: financial, programmatic, procurement, and market take-up.

In particular, in the financial area, the main benefits are improved GALILEO business case based on integrated institutional and commercial structures and service provision mitigating competition risk; early market development and revenues from EGNOS; improved private-sector investor confidence from continued public-sector support of EGNOS; and improved cost control as the EGNOS experience might mitigate some technical risks.

In the programmatic area, the benefits will accrue from increased likelihood that GALILEO OS and SoL (see Chapter 5) service will be introduced in 2008 and 2010, respectively, due to: a successful EGNOS safety-case mitigating GALILEO safety-case risks; EGNOS provision of an early opportunity to develop standards and procedures for safety-related applications; and EGNOS development experience mitigating some technical risks.

In the procurement area, the main benefit is the improved business case of Concessionaire bidders based on: early revenues from EGNOS; continued public-sector support to EGNOS having a positive impact on private-sector investor confidence; integrated service provision mitigating competition risks; and optimized public/private sector risk allocation from optimal integration of EGNOS into GALILEO.

Finally, the market take-up benefit is mainly based on: EGNOS increasing market awareness of GALILEO; a successful EGNOS safety-case mitigating risks linked with GALILEO safety-case; and early development of standards and procedures mitigating risks associated with GALILEO SoL services.

In addition to the expected revenue potential as a result of the integration of EGNOS into GALILEO, the exploitation of EGNOS to prepare markets for GALILEO services outside Europe, in competition with other augmentation systems (see Chapter 4), should improve the GALILEO business case and ensure non-European markets are still open to GALILEO when it starts operation in 2008. Analysis of the integration options for EGNOS and GALILEO shows that operating cost savings could be achieved: savings due to integration could represent up to 9% of the combined operation costs in case the two systems remain functionally independent and up to 12% if they are fully integrated [19]. It is not likely that any capital cost savings can be obtained as a result of the integration of the two systems; however, integration also could be beneficial to EGNOS itself in reducing costs during the 2004–2009 period.

7.4 GALILEO

Chapter 5 presented a detailed description of the GALILEO system. Due to the fact that GALILEO is under development, a detailed description of the program structure, scheduling, and organization was provided to support, integrate, and make the technical information effective. Therefore, in what follows, some key aspects—more directly related to the legal and market issues—that were omitted in Chapter 5 are provided. Hence, a good comprehension of the programmatic issues covered in Chapter 5 is a prerequisite for the material covered in this section.

7.4.1 Legal Framework for GALILEO

The GALILEO system will operate in a very complex legal and institutional environment, accounting for the number of services to be provided and users to be served across the world, particularly in the area of major safety and security-related applications. Therefore, in the frame of a GALILEO study (Galilei Project), a legal team has analyzed the status of the law relevant to GALILEO, evaluating where problems, gaps, and overlaps arise and issuing recommendations for dealing with them [23]. A legal/functional model was established to allow a high-level mapping of relevant legal issues to the key players (Figure 7.5) [23]. Moving from that model, analysis of the legal regimes applicable to the various sectors becomes possible, at both the national and European and even global levels, as well as analysis of their impact for the delivery of GALILEO signals and services.

The GALILEO institutional framework consists of both a public monitoring entity, the GALILEO Supervisory Authority (see Chapter 5), and a private operator, the GALILEO Operating Company (see Chapter 5), linked by a Concession Agreement. Subsequently, a GALILEO Convention could be established, as a legal coverage of the Concession. The major legal relationship of the GALILEO core system is expected between providers of value-added services and end users. This includes the regional components, providing regional integrity, and local elements, augmenting the GALILEO global services by either enhancing the core signals or offering value-added services based on the GALILEO signals (see Chapter 5).

Contractual agreement may exist between the GOC and third-party service providers in order, for instance, to provide service guarantees referring to levels of offered service (CS), integrity guarantees (SoL), or security guarantees (PRS), together with the liability arrangements that derives from payments of those services. The Concession Agreement needs to regulate the main parameters that impact on the GOC commercial freedom to arrange such contracts.

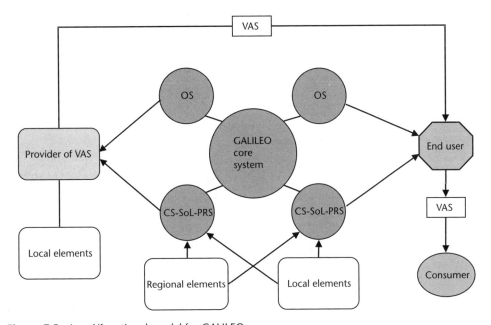

Figure 7.5 Legal/functional model for GALILEO.

The Concession Agreement has to deal in detail with key legal issues related to the GALILEO business frame, such as the sharing of risk within the PPP (see Chapter 5), various long-term planning issues, as well as the respective mechanisms available for financing the system and generating revenues from its operations.

Since GALILEO will be involved in a number of crucial safety- and/or security-sensitive operations, liability in case of damage is a major issue that has to be dealt with in the Concession Agreement and in the contracts of the GOC with its customers. On the other hand, the GOC will not produce and sell receivers: as a consequence, product liability claims and third-party liability claims under noncontractual liability are unlikely to be brought against it. GALILEO services are not likely to cause damage: the major risks lie, instead, in the downstream. Nevertheless, it is proposed that the GOC offers broad liability coverage through the contract chain, as depicted in Figure 7.6 [23].

The GOC should offer a broad derogation of liability to its customers, in case they are involved in contractual and noncontractual liability claims, hence ensuring the GOC against a first tier of liability compensation as agreed under the contract; a second tier of liability would then be compensated through the Concession and a compensation fund. Value-added services or end users will be covered by existing liability regimes of the contractual chain and will have access to claims if they can prove GALILEO services to be the ultimate cause of damage.

Further important results of the Galilei legal task concern:

- Patents that should be developed on the GALILEO SIS for the production of compatible receivers by the establishment of patent pools with manufacturers interested in aligning themselves with the GALILEO trademarks. These can provide a major source of revenue from the OS.

- A policy on access control through encryption and service denial that ensures, in particular with respect to the PRS, a security level matching Europe's requirements.

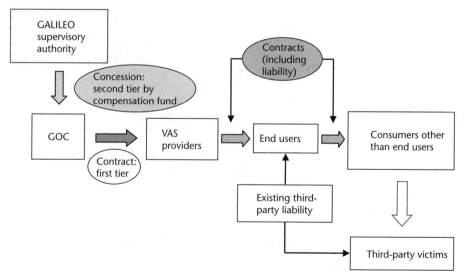

Figure 7.6   GALILEO liability and the contractual chain.

- Important enhancements provided by GALILEO to the existing search-and-rescue satellite system without legal constraints, acting within the relevant legal framework.

As far as security issues are concerned, once GALILEO is operational, European policy-makers will have the option of using its signals to boost the scope of both the Common Foreign and Security Policy (CFSP) and the European Security and Defense Policy (ESDP). For development of the Rapid Reaction Force (RRF), access to GALILEO would serve to enhance its operational performance [24]. At a political level, an independent navigation and positioning system increases the leverage of EU CFSP; on the other hand, ownership of a GNSS brings added responsibilities. While the European Commission considers GALILEO a civilian system, other countries may not. At an operational level, this implies that European policy-makers need to be prepared to deal with the eventual unauthorized use of the system by third parties.

A permanent forum/working group is needed between the EU and United States to handle ongoing and future outstanding issues, composed of decision makers of comparable levels with the appropriate clearance. An additional working group, composed of military representatives from both sides, should be considered to coordinate transatlantic policies in time of crisis (i.e., asymmetric signal uses and jamming).

7.4.2   The Joint Undertaking Statutes

The innovative approach that led to the conception, development, and realization of the Joint Undertaking has been stressed in Chapters 1 and 5, to enable the reader to become familiar with the basis of the GALILEO challenges. What follows provides a more legal view of the JU by an "active reading" of its Statutes, composed of 23 Articles [25].

Location (Brussels) and composition (EC, ESA as founding members) is regulated by Article 1, where the possibility to become members is extended to the European Investment Bank and any undertaking that has undergone the envisaged tendering procedures, as outlined in Chapter 1. In addition, Article 1 regulates the composition of the JU capital, including the assets brought in by its members; founding members share a capital to the extent of the amounts indicated in their commitments of 250 million euros for EC and 50 million euros for ESA, whereas undertakings need to subscribe 5 million euros (large enterprises) and 250,000 euros (small/medium enterprises, or SMEs) per year. Under Article 1, the JU can make financial commitments that do not exceed the available amount of capital.

Article 2 is crucial, as it states the main tasks of JU, in particular: handling of the optimal integration of EGNOS in the GALILEO program and implementation of the GALILEO development and validation phases as well as support in preparing the deployment and operational phase; launch, according to the regulated procedures, of the research and development activities needed to complete the program successfully, including the launch (through ESA and in accordance with the regulated procedures) of a first series of satellites to enable an end-to-end demonstration of system capabilities and reliability; aid in mobilizing public- and private-sector funds necessary to issue proposals to the council for the management structures of the different phases.

Articles 3 (recalled twice in Article 2) and 4 regulate the JU agreement with ESA and contract for the provision of services, respectively. The JU legal personality is stated in Article 5, which underlines, in particular, its capability of acquiring or disposing movable or immovable property and of being a party to legal proceedings. Article 6, instead, regulates the JU property in the form of tangible and intangible assets created or transferred to it for the GALILEO development phase.

The type, composition, and duties of the JU bodies(i.e., administrative board, executive committee, and director) are regulated by Articles 7–10, whereas Article 11 regulates the staff complement.

The financial aspects related to revenues, financial year extension, cost estimates, financial regulations, annual accounts, and balance sheets are dealt with through Articles 12–15.

Article 16 regulates the program development plan and the annual report, whereas liability and insurance issues are covered in Article 17.

Protection of sensitive information is regulated by Article 18, where it is required to take the experience of the GSSB into account.

Access to the JU by members different from those referred to in Article 1 is regulated by Article 19.

Article 20 states the JU should operate for four years, with possible extension by properly regulated amendment of the Statutes. Liquidators are regulated by Article 21, whereas Article 22 underlines that any matter not covered by the Statutes has to be considered in accordance with the law of the State where the JU seat is located. Finally, Article 23 regulates the process for submitting a proposal by any JU member to amend the Statutes.

7.4.3 Supervisory Authority and Centre for Security and Safety Regulations

As outlined in Chapter 5, the European satellite radio-navigation system has a strategic nature and the necessity to adequately defend and represent the essential public interests. In this context, the need to entrust to a public authority the supervision of the deployment and operational phases of the system resulted in conceiving the Supervisory Authority and the Centre for Security and Safety [26]. Chapter 5 highlighted the main tasks of the two structures and framed this within the deployment process of the GALILEO program.

As far as further legal aspects are concerned, a regulation of 23 articles handles the Supervisory Authority and the Centre for Security and Safety activities. In particular, the 19 articles devoted to the Supervisory Authority cover its aim and objective (Art. 1), tasks (Art. 2), ownership (Art. 3), legal status, seat, and local offices (Art. 4), administrative board (Art. 5) and its tasks (Art. 6), director (Art. 7) and his or her tasks (Art. 8), Scientific and Technical Committee (Art. 9), budget (Art. 10), its establishment (Art. 11), implementation and control (Art. 12), the financial regulation (Art. 13), antifraud measures (Art. 14), privileges and immunities (Art. 15), staff (Art. 16), liability (Art. 17), access to documents and protection of data of a personal character (Art. 18), and participation of third countries (Art. 19).

It is worth mentioning that these regulations bestow a legal status to the Supervisory Authority, enabling it to act as a legal person in the discharge of its tasks.

The four articles regulating the Centre for Security and Safety cover the setting-up of the center (Art. 1) and tasks (Art. 2), its composition and operation (Art. 3), and the entry of force (Art. 4).

7.4.4 The PPP Bidding Process

In the frame of Phase II of the GALILEO Inception Study an interesting document was issued [19] to support the development of a business plan for the navigation program. The topics addressed here provide a flavor of the legal and market-related issues relevant to the actual deployment of a complex engineering structure such as the GALILEO system: the procurement plan for the GALILEO PPP; intellectual property rights (IPR), in particular the potential for generating revenues by IPR protection of the GALILEO chipsets through royalties from licensing the technology; and the optimal way to proceed with the European augmentation approach [i.e., EGNOS (see Section 3.2)] in relation with the GALILEO program.

The previous phase (Phase I) of the Inception Study offered a thorough review of services that should be offered and revenue that could be generated, as well as specifications and costs, case for public- and private-sector investments, structure for a PPP, and a strategy for procuring and financing the system. This review demonstrated that there was a case for public-sector support for GALILEO and that the private sector would be willing to participate in certain circumstances in a PPP. The PPP should be implemented by awarding a Concession. Under this approach, the JU, which manages the Concession award process and the development phase, subsequently awards a Concession for the deployment and operation of GALILEO to a private-sector consortium through a competitive tender.

A clear separation between functions and responsibilities of the private and public sectors during the deployment and operations phases is vital to allow the public and private sectors to concentrate on what they do best. The conceived Concessionaire model is expected to generate revenue from royalties on chipset sales, which are paid by equipment providers incorporating a GALILEO chip in their products to allow users to receive the open access service, and from service providers, which use a specialized encrypted signal to offer other value-added services based on the integrity service, the commercial service, and improved service performances (see Chapter 5).

The main objective of the PPP is the achievement of value of money for the public sector by transferring proper risk and responsibility to the private sector through an effective approach that creates incentives to optimize the public-sector benefits. In particular, PPP aims at procurement: efficiency optimization by giving the private sector responsibility for ensuring that system performance and specifications meet the performance requirements of both the commercial market and the public sector; optimization of the revenue generation from the market; reduction in the need for public expenditure and ability to spread the public contribution over a longer period; and optimization of overall life costs by introducingprivate-sector efficiencies.

The Procurement Plan includes three phases of the bidding process: initial bids with the selection of a limited number of bidders, subsequently invited to develop full concession bids. The restriction on the number of bidders in the final concession bid phase is aimed at minimizing costs and providing an acceptable risk-reward ratio for the bidders. The criteria for the initial bid are mainly qualitative, whereas the concession bid evaluation criteria envisages the possible financial implications for the public sector, including revenue/profit sharing proposals, innovation in service provision, credibility and potential of the achievement of the objectives of the proposed business plan, and underlying evidence supporting the business plan (e.g., offers for system supply, offers of financing, market research, and any agreements or memoranda of understanding with potential customers and partners).

The Procurement Plan was conceived to achieve the concession bids in advance of both the system critical design review (CDR) and in orbit validation review (IOVR) of the development phase. In case any design change would be required as a result of these reviews, there would be a suitable mechanism along with a financial adjustment between the public sector and the Concessionaire. Competencies of the latter include procurement and operation of sophisticated satellite systems and market development for exploitation of navigation products and services embodying new technology and structuring complex financial packages.

The GALILEO program benefits from a public-sector comparison activity, aimed at quantifying the costs and revenue-generating potential of undertaking the whole project under public-sector control. This provides a benchmark to measure the overall benefit of private-sector bids and, as a consequence, improve the public-sector negotiating position. In addition, the comparison enables a detailed analysis of risks and their potential impact on costs and revenues.

The bidders competition is a key step for achieving value of money: credibility and capability of procurement in a cost-effective manner are keys of success in this respect. In the frame of GALILEO, achieving those goals is particularly challenging, considering that the development phase, under the ESA technical management, is crucial for developing the necessary technology within Europe, and the difficulty of forecasting revenues on a nascent market, as is the case with satellite-based navigation services.

Some important features and implications of the development phase are highlighted next. First, possible partial competitive procurement for the deployment phase satellites from outside Europe could be possible, given the technical know-how related to satellite navigation systems already possessed by other non-European countries. Second, the Concessionaire ability to raise finances for the acquisition of the GALILEO system in the debt market will be crucial. Third, about 60% of the ground segment cost will be incurred in the development phase, mostly in terms of software costs. The latter needs to be rendered available to the Concessionaire by the JU and the contractual arrangements in the development phase have to ensure this is possible. The ground segment, to be built in the deployment phase, includes off-the-shelf equipment and specialized buildings as well as software. The equipment will include commodity items, where the ability to generate value for money for the public sector will be limited. The ability of the private sector to generate value for money improvements, which will mostly be related to specialized buildings, is expected to be limited. However, this is not a major concern as the ground segment only represents about 20% of the Concessionaire capital investment.

In the development of the Concession agreement between the JU and the Concessionaire to ensure that its operations, as the publicly mandated supplier of navigation services in Europe, are conducted in the public interest, the following has to be considered: obligations to obtain and maintain licenses; access by service providers to ensure consumer choice among different suppliers; universal service and coverage/rollout obligations; pricing controls in relation to any service where the Concessionaire is a dominant supplier; publication and disclosure of data to provide a high degree of transparency for consumers and regulator; cooperation with the GALILEO regulatory body and sector-specific regulatory bodies; powers of audit and inspection; and safety, liability, and security issues.

7.4.5 Toward Business Development for the European Industry

Traditional European industry strengths have been shown in chipset design, system integration, and service provision. However, the acquisition of the Ashtec and Magellan brands by Thales has seen European firms take a stronger hold of the manufacturing segment, even if most production remains in the Eastern countries. In terms of market share, the key sector seems to be personal mobility (Figure. 7.7) [23].

The low domestic demand for GPS-enabled cell phones indicates that none of the main European manufacturers currently has any significant presence in the GPS-enabled handset market [23].

The global revenue for GNSS products and services is estimated to increase by 250% between 2010 and 2020. Opportunities for European industry lie in capturing at least a third of these worldwide revenues.

Building on these market share figures, the gross revenues that will accrue for European suppliers from the global markets in 2015 have been estimated [23]. Some results are provided in Table 7.3, which shows how the largest part of the revenue (58%) is at the product level (i.e., in system integration).

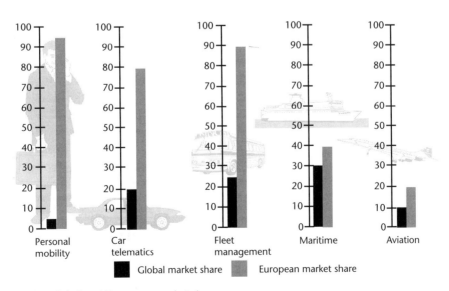

Figure 7.7 Global and European market share.

Table 7.3 European Players' Revenue by Value Chain Segment (2015)

Segment	Share (billions of euros)
System integration	47.3
Service	34.3
Component	1

European competitiveness in the GNSS global markets can be increased by focusing on the strengths of the European players and investing in market segments and application areas with the highest profit and growth potential. The major opportunities for European players lie within the system integration segment for application markets, such as telematics, personal mobility, and surveying, whereas at the component level, only specialized and niche products should be considered.

7.4.6 Standards and Certification

GALILEO, as the newest satellite positioning and navigation system, will have to comply with all necessary, existing international standards and certification requirements [23]. This is particularly important for the SoL service.

Some key activities in the standardization domain include: promotion of the SoL standards jointly with the Setting Standards for GALILEO (SAGA) EC project covering all standardization issues; contributions to technical discussion within main standardization bodies such as ICAO, IMO, the European Organization for Civil Aviation Electronics (EUROCAE), and the Requirements and Technical Concepts for Aviation (RTCA); development of a concept for operational use of GALILEO and GALILEO+GPS receivers with optimum interoperability; radio frequency interference mitigation techniques and associated standards for future SoL receivers; and assessment of GALILEO services versus SoL applications in the aviation and maritime domains.

In order to be considered and used globally, GALILEO should be clearly identified in the standards of the Third Generation Partnership Project (3GPP)—a collaborative agreement between a number of telecommunications standard bodies and market representation partners such as equipment manufacturers and mobile communications operators, with the aim of producing globally applicable technical specifications for 3G mobile systems—that are being established by the mobile phone community. In this context, promoting GALILEO in the 3GPP community and identifying the impacts on current standards of including GALILEO as an additional GNSS solution for location-based services have been performed. The required level of confidence in the GALILEO signals, including their accuracy, availability, integrity, and compatibility with application certification requirements, can be achieved through the design and implementation of a GALILEO certification procedure. The latter also represents a milestone for future application certifications. The creation of a certification authority, linked to the EC, is a key step to defining the certification requirements and notifying and controlling a certification body based on advice from experts. The body is expected to provide written assurance that the system's development, implementation, operation, and dependability conform to specified requirements. The certification approach is depicted in Figure 7.8 [23].

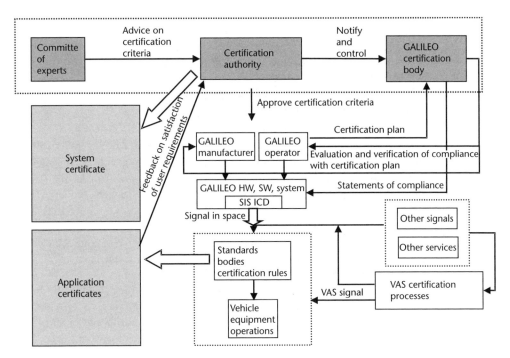

Figure 7.8 Certification approach for GALILEO.

References

[1]   *Business in Satellite Navigation—An Overview of Market Developments and Emerging Applications,* Rep. GALILEO Joint Undertaking, March 5, 2003.

[2]   Spada, M., "Aeronavigazione Satellitare e Eommercializzazione Nello Spazio," *A. Giuffrè Editore,* Milan, Italy, 2001.

[3]   "2001 Federal Radionavigation Plan," Department of Defense, Department of Transportation, 2001.

[4]   Kaplan, E. D., (ed.), *Understanding GPS: Principles and Applications,* Norwood, MA: Artech House, 1996, Chapter 12, pp. 487–516.

[5]   Rycroft, M., (ed.), "Policy, Commercial and Technical Interaction," Kluwer Academic Publishers, *Proceedings of an International Symposium,* Strasbourg, France, May 26–28, 2003.

[6]   "GPS World Markets: Opportunities for Equipment and IC Suppliers," ABI Researcher, 2003.

[7]   Gurtner, W., "RINEX: The Receiver-Independent Exchange Format," *GPS World,* Vol. 5, July 1994, pp. 49–52.

[8]   Gurtner, W., and L. Estey, "RINEX Version 2.20: Modification to Accommodate Low Earth Orbiter Data," University of Berna, April 12, 2001, ftp://ftp.unibe.ch/aiub/rinex/rnx_les.txt.

[9]   Remondi, B. W., "Distribution of Global Positioning System Ephemerides by the National Geodetic Survey," *First Conference on Civil Applications of GPS,* Institute of Navigation, September 1985.

[10] Remondi, B. W., *Extending the National Geodetic Survey Geodetic Orbit Formats,* NOAA Technical Report 133 NGS 46, 1989.

[11] "RTCM Recommended Standards for Differential GNSS Service," Radio Technical Commission for Maritime Services, Version 2.2, Alexandria, VA, January 1998.

[12] "NMEA 0183 Standards for Interfacing Marine Electronics," National Marine Electronic Association, Version 3.0, NC, July 2000.

[13] *2001 Federal Radionavigation Systems*, Department of Defense, Department of Transportation, Final Report, November 2001.

[14] "WAAS Benefits Register," April 23, 2004; http://www.gps.faa.gov/Library/index.htm.

[15] http://www.thomas.loc.gov/home/approp/app04.html.

[16] http://www.faa.gov/aba/html_budget.

[17] Weber, D., "Happenings on the Hill," *SatNav News*, Vol. 21, November 2003.

[18] Sigler, E., "WAAS International Expansion," *SatNav News,* Vol. 20, June 2003.

[19] "Inception Study to Support the Development of a Business Plan for the GALILEO Programme," Executive Summary Phase II, PriceWaterhouseCoopers, January 2003.

[20] "GALILEO—Integration of EGNOS—Council Conclusions," n. 9698/03 (Presse 146), June 5, 2003.

[21] "Integration of the EGNOS Programme in the GALILEO Programme," Communication from the Commission to the European Parliament and the Council, COM(2003) 123 Final, March 19, 2003.

[22] "The Commission Proposes Integrating the EGNOS and GALILEO Programmes," IP/03/417, Brussels, Belgium, March 20, 2003.

[23] "The Galilei Project—GALILEO Design Consolidation," European Commission, ESYS Plc, August 2003.

[24] Lindstrom, G., and G. Gasparini, "The GALILEO Satellite System and Its Security Implications," EU Institute for Security Studies, Occasional Paper No. 44, April 2003.

[25] "Statutes of the GALILEO Joint Undertaking," Official Journal of the European Communities, May 28, 2002, L.138/4–8.

[26] Proposal of "Council Regulation on the Establishment of Structures for the Management of the European Satellite Radionavigation Programme," n. COM (2003) 471 Final, 2003/0177 (CNS), July 31, 2003.

CHAPTER 8

Layer Issues

8.1 Introduction

The previous chapters showed the complex architecture, features, and services of the first, second, and intermediate generation of satellite navigation systems. The most fascinating feature of GNSS is their potential to cover a wide range of applications.

Both current and future applications of GNSS can be framed in a height-dependent model, where users are located in three main layers:

1. *Land/water (L/W) layer* (Earth surface, oceans/seas/lakes/rivers);
2. *Air (A) layer* (atmospheric region);
3. *Space (S) layer* (above atmospheric region to deep space).

Each of the three layers hosts systems for various applications (e.g., road/rail transportation, navigation, aviation, surveying, leisure) that are or can become users of the GNSS services to improve performance/effectiveness/cost of the systems. A view of the layer model is exemplified in Figure 8.1, where some categories of users are pointed for each layer.

The following sections present emerging and potential applications of GNSS grouped and described layer by layer.

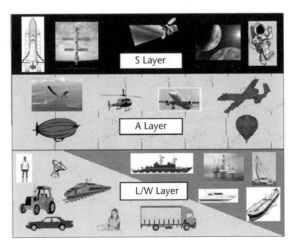

Figure 8.1 The layer model for GNSS users and applications.

8.2 Land/Water Layer

8.2.1 Land

8.2.1.1 Intelligent Transportation Systems

Intelligent transportation systems (ITS), also meant as *transport telematics systems,* are electronic, information, and communication technologies and traffic engineering concepts used in an integrated way to better manage and improve the performance, efficiency, and safety of the transportation system. Since 1995, the ITS America and Department of Transportation have started to provide a clear vision of what industry and government should do about ITS [1].

GPS was a key technology for developing advanced ITS services. This was followed by the development of a coherent policy by government authorities [2, 3]. It is important to understand as well that the correlation among different user services for ITS operations is very tight. This means that some services can share the same physical, technological, and information infrastructure to reduce costs and improve performance. The first step was to define a service bundle in which each category of service can be grouped. Although not exhaustive, the bundle helps to segment the possible application of ITS. The list of the original 29 user services has been extended to the 33 ITS user services (Table 8.1) [3]. This list will change over time as user needs change and increase and new (and improved) technologies emerge. It is beyond the scope of this book to detail the ITS architectures and planning [4], but the role of satellite navigation on ITS will be explained through the example of developed applications and future trends. GPS is mainly used to locate single vehicles or fleets. The positioning information enables a wide range of services, depending on their interaction with digital mapping, control centers, vehicle sensors, and other users.

Route Guidance

The interaction of the position with digital mapping provides the user with route guidance. This service enables the user to know street-by-street the route to follow to reach a selected destination, via visual and audio interfaces. Route guidance can be performed with several user terminals with different performances. The terminals are commonly referred to as "satellite navigators." User terminals can be integrated with vehicular motion sensors equipment if required. The main sensors for vehicular navigation are the odometer and gyroscope. The sensor data can be processed by the satellite navigator to compute a positioning solution, without the help of GPS. This methodology is known as *deduced reckoning* (DR) and the systems that calculate the solution are called *inertial navigation systems*. The main drawback of an INS when operated as a stand-alone system is the growth of systematic errors with time. An INS can operate in synergy with the GPS; therefore, an INS is able to recalibrate its solution to the GPS every time a GPS solution is available [5]. An INS improves the satellite navigator performance when a GPS solution is not available because of impaired environmental conditions (i.e., urban canyons, tunnels, thick surrounding vegetation); in these circumstances, the satellite navigator uses the less accurate but always available INS positioning solutions. The technological trend is to integrate in the same user terminal many of the car entertainment functions, such as FM radio, CD player, DVD player, mobile phone, TV, and

Table 8.1 User Services

User Service Bundle		User Service
Travel and Traffic Management	1.1	Pre-trip travel information
	1.2	En route driver information
	1.3	Route guidance
	1.4	Ride matching and reservation
	1.5	Traveler services information
	1.6	Traffic control
	1.7	Incident management
	1.8	Travel demand management
	1.9	Emission testing and mitigation
	1.10	Highway-rail intersection
Public Transportation Management	2.1	Public transportation management
	2.2	En route transit information
	2.3	Personalized public transit
	2.4	Public travel security
Electronic Payment	3.1	Electronic payment services
Commercial Vehicle Operations	4.1	Commercial vehicle electronic clearance
	4.2	Automated roadside safety inspection
	4.3	Onboard safety and security monitoring
	4.4	Commercial vehicle administrative processes
	4.5	Hazardous material security and incident response
	4.6	Freight mobility
Emergency Management	5.1	Emergency notification and personal security
	5.2	Emergency vehicle management
	5.3	Disaster response and evacuation
Advanced Vehicle Safety	6.1	Longitudinal collision avoidance
	6.2	Lateral collision avoidance
	6.3	Intersection collision avoidance
	6.4	Vision enhancement for crash avoidance
	6.5	Safety readiness
	6.6	Precrash restraint deployment
	6.7	Automated vehicle operation
Information Management	7.1	Archived data function
Maintenance and Construction Management	8.1	Maintenance and construction operations

Source: [3].

infomobility services; an example of one of these advanced user terminals is shown in Figure 8.2.

Some user terminals do not use the vehicular sensors. These are usually the portable ones. This class includes personal digital assistants (PDAs) and laptops integrated with GPS receiver and digital mapping software, using standard operative systems (i.e., Windows, Palm, Symbian). Most satellite navigators can use the signals from SBAS (i.e., EGNOS/WAAS) to improve performance. The performance of

Figure 8.2   Example of advanced user terminal for infomobility.

a route guidance system is also strictly correlated with the accuracy and availability of the digital mapping. The route guidance can be dynamically influenced by the information provided by a wireless network. This includes traffic jams or strategic decisions taken by a traffic management system (e.g., avoiding a road in which there is a demonstration, road blocking, and so on). The information is often transmitted by a radio channel or by cellular mobile networks. In Europe the traffic message channel (TMC) is becoming a "de facto" standard for traffic information.

Traveler Services Information (Infomobility)

The position of the traveler is the key information to enable infomobility services. Infomobility is the set of technologies and applications that allow the "user on the move" to access geographic and location-based information. The implication of satellite navigation on traveler services information in the frame of ITS is discussed in this section, whereas location-based services will be discussed in Section 8.2.1.2. The positioning information of the traveler can be sent to the control center at the moment a service is requested or updated periodically (tracking mode). The most common means for communication is the mobile cellular network; the information can be provided by voice call, short message service (SMS), or data transmission (e.g., GPRS). It is likely that infomobility services will be strongly influenced by the evolution of 3G and 4G mobile networks [6]. The most common services are related to the safety and efficiency of the journey, such as weather forecasts, route suggestions, traffic or road blocking information, road conditions, heavy or hazardous cargo transportation, and gas station locations. Each of these services can be provided "on request" or as a "pop-up" at the moment of reception of an event. The position and planned route enable the central control to provide only the information of interest to the traveler. In general, terminals used to provide infomobility services integrate car entertainment and route guidance functions, whereas only some of them have a vehicle antitheft function. The vehicle security service is supported by integrating the antitheft sensor of the vehicle (i.e., movement sensors, vibration sensors) with satellite navigation and mobile communication. If the vehicle sensors verify the event of theft, an automatic message is sent to the control center. Tracking information informs the user and public authorities about the vehicle position and movement. Many user terminals for advanced infomobility services have been developed, but the mass market has not adopted them yet. In addition, the costs associated with providing these services are also relatively high because of the small number of user terminals on the road. Each user terminal has proprietary standards for communication and proprietary technologies and this does not help decrease the costs. Market analysis is very optimistic with respect to the foreseen revenue of these services, even if the road to decrease the prices looks far from smooth. Infomobility needs standards and there are three possible solutions for this: convergence of major user terminals manufacturers toward an agreed common standard (for hardware, software, and communication protocols), even with governmental policy help; convergence of minor manufacturers toward a standard imposed by a major manufacturer; convergence of all manufacturers toward personal computer (PC) hardware and software protocols. The latter seems to accomplish the wishes of the users; in fact, it is usual to find self-made user terminals (mainly for car entertainment) in the vehicles based on PCs. The advantages of these custom-made terminals are their ability to use a great software library working under standardized operative systems (e.g., Windows, Linux); their state-of-the-art and low-cost hardware; their ability to be easily upgraded; their ability to use Internet information services; and their cost-effectiveness.

Traffic Control (Probe Cars)

Traffic flow measurement is one of the most challenging problems for ITS. The technologies for traffic flow monitoring are many and their performance can be excellent if well integrated. The main devices for traffic flow measurement are closed-circuit TV cameras (often integrated with a video image processor), red light cameras, microwave vehicle detection systems (MVDS), infrared detectors, acoustic and ultrasonic detectors, inductive loop detectors, and ramp metering sensors. Nevertheless each of these technologies suffers drawbacks such as high equipment costs, high installation or maintenance costs, low performance under certain conditions, and very short range or short life. Satellite navigation can give the decisive aid, because each user terminal installed on a vehicle can record and transmit (by mobile network) its real-time position, direction, and velocity. These parameters, if integrated with similar information from a wide number of user terminals, can give accurate and wide area information on traffic flow [7]. It is likely that the use of this technique will significantly increase over the years, especially when advanced and standardized user terminals for infomobility services with two-way communication capability has a mass-market impact. Traffic flow measurement could be a no-cost feature for this class of user terminals.

Accident Management

Satellite navigation can be of fundamental importance to managing vehicle collisions. The detection of road accidents is usually left to nonautomatic systems (e.g., spontaneous emergency calls). The accident can be detected by a video camera only if the scene is under the coverage of a video-surveillance system. This is not the case for most road networks. The integration of in-vehicle sensors, satellite navigation, and mobile communications can enable a safety-of-life incident detection system. Sensors on cars (e.g., crash sensors based on accelerometers) can detect crash events and the communication means can be used to send the last calculated position to a control center. The control center can decide to send emergency vehicles (police, ambulance, and so on) depending on the sensor response (i.e., the seriousness of the accident), the type of road, the number of cars involved, and voice call data from the driver or passenger. The service's effectiveness is clear in difficult situations, such as: a road in a rural area and vehicle passengers unable to call for emergency help; nighttime accident in low visibility for other drivers or harsh terrain (e.g., down an escarpment), and vehicle passengers unable to call for emergency help; every situation in which the seriousness of the accident needs automatic detection systems for immediate rescue operations. In the event of a road accident, the control center can send an immediate emergency communication to all vehicles nearby with the same user terminal. The information about an accident close to the user position can be used for first-aid purposes and to avoid chains of accidents in the same area. The user terminal can record all sensor data (including position, speed, accelerations) for postaccident analysis (i.e., black box). When integrated with so-called "panic buttons," the user terminal could be used in case of an emergency (i.e., car failures, threats, need for help). In this case the user position is transmitted to the control center to provide the service requested.

Public Transportation Operations

The real-time knowledge of the vehicle position by satellite navigation systems enables advanced public transportation management and other services. The actual journeys can be managed by a control center and compared with the schedules. The schedule can be changed or modified "in near real time" if traffic conditions delay the vehicles. Information about the position of vehicles (and the expected arrival time) can be communicated to the user by Internet, mobile phone (SMS), or via variable message signs (VMS) at bus stops. Furthermore, the *personalized public transit* (PPT) service can benefit from satellite navigation systems. There are basically two types of PPT: flexibly routed operations and random route operations. Both types try to satisfy the demand of mobility by flexible vehicle routes based on the requests received. The goal of PPT is to implement a near "door-to-door service." The knowledge of the position and scheduled route of these "on-demand" vehicles can enable the user to make more efficient decisions on which path to travel along. Nevertheless the user terminal can integrate communication and mapping (e.g., hybridized mobile phone or PDA), inform about the time and best site to catch a bus to reach a selected destination, and even reserve the stop. The control center can register the new reservation and send this information to the vehicle's navigation system. The driver will follow the route guidance of the navigation system, without any need for voice communications with the control center. An interesting application of GPS is the taxi dispatching service [8]. This is a clear extension of the PPT concept, where the role of choice and coordination is held by the control center that decides which vehicle should take the job, depending on the vehicle and customer locations.

Electronic Payment Services

Electronic payment services (EPS) refer to all the systems that provide travelers with a common electronic payment medium for different transportation modes and functions. The first generation of EPS mainly uses electronic cards (i.e., prepaid magnetic strip cards or contact-less smart cards) for common fare payment (e.g., highway tolling, parking fare, or public transport ticketing). The common denominator of this class of EPS systems is that the user has to interact with the electronic system; paying with the mobile phone using SMS is another example. Prepaid electronic cards are still widespread for tolling payments even if the wide use of credit cards is gradually rendering this system obsolete.

The second generation of EPS foresees no interaction between the system and user. For example, the system allows drivers to pay highway tolls without stopping at the barrier. It works using a specified user terminal that uses short-range communications to identify the user, or by using closed-circuit cameras that automatically read the number plates of in-transit vehicles. Bluetooth payment terminals for parking fares and public transport ticketing are included in this class, because the system uses wireless communications without the need for user action.

The future trend foresees the third generation of EPS in which the system has the knowledge of the user position and, hence, can provide advanced services. Satellite navigation is a key tool for this new EPS. The first application of this concept is the remote tolling service, using a user terminal with positioning and communication capabilities. The toll is paid when the position measured by GPS (eventually with the help of SBAS) places the vehicle in a road where the toll must be paid. The system can also be used for control of access to restricted areas (e.g., city centers), the search for a parking space and payment, and to detect traffic offenses. The European Commission recently established by a directive [9] that future electronic toll collection systems have to be interoperable and/or commonly adopted by all members of the European community. The GALILEO JU (Joint Understanding) recently cofunded a project called *vehicular remote tolling* (VeRT) to study and demonstrate this new EPS concept using EGNOS and mobile communications [10]. The extension of this generation of services for nonvehicular users is still to be studied, because there is not a reliable, efficient, and low-cost sole means for personal localization in urban and indoor areas.

However, it is important to note that even the first generation of EPS has not achieved worldwide application yet.

Other Services

There are several niche services in the frame of ITS that can be enabled by using satellite navigation. This section does not want to be exhaustive or depict future trends about niche services; instead it aims to give readers a better understanding that even the most traditional application can be modernized with the help of positioning information. On the other hand, some advanced concepts about ITS are becoming a reality and should revolutionize (in the long term) our common idea of mobility.

The traditional insurance service for vehicles can be modernized in a new concept of *pay-per-use insurance*. The user position is calculated by satellite navigation systems and communicated to the control center by mobile communication networks and some important parameters are recorded in a "black box" device. An insurance control center can track the movement of the vehicle to evaluate the insurance cost to the user, as a weighed function of different parameters. Some of the most important parameters are the time percentage of use, but also the exact time intervals of use in the insurance reference period; the typology of the environment in which the car is normally used; onboard information that can affect the final insurance cost; and the "black box" data, in case of an accident.

The *environmental parameters* are becoming more and more important for our day-by-day life, for our health, and for the future of our planet. Local governments of major cities usually have the authority to manage the local environment (mostly in terms of pollution, temperature, humidity, pressure monitoring) and use the relevant data to develop local policies for reducing traffic congestion. The environmental monitoring is normally carried out by measuring the levels of a number of parameters using specialized sensors (pollution meters) installed at some "key points" within a city. These devices can give only a local measurement and, therefore, cannot be used to understand wide area impacts. Special vehicles equipped with environmental sensors could be used to measure the relevant parameters giving the exact time and position of the measurement. The records could be sent in near-real time to a control center and/or stored aboard for a postmission analysis.

The real-time knowledge of the vehicle position is of fundamental importance both for hazardous and valuable goods tracking and operation schedule. The themes of signal integrity and communications security are of fundamental importance for these applications. Nevertheless the major forwarding agencies use satellite navigation technology to track vehicles and give the user the service of goods tracking. A real example of a special vehicle carrying both navigation equipment and environmental sensors is the *vehicle performance and emissions monitoring system* (VPEMS), whose implementation and test phase are presented in [11]. The aim of the project behind this vehicle is the realization of a system able to collect self-consistent traffic, travel, and environmental data to develop the necessary scientific bases for an effective environment policy. Conventional means for gathering city pollution, such as roadside environmental monitors, travel surveys, and so on, cannot provide data on individual vehicles and on space-temporal aspects of individual drivers' travel behavior because they measure average concentrations, not emissions; in addition, many health impacts are related to cumulative exposure. From this set of motivations, the VPEMS project had started its development. The designed vehicle was conceived to capture simultaneously and manage space-temporally referenced travel, traffic, and environmental information through many means, namely, GPS, GIS, GSM, data warehousing and data mining, and vehicle-embedded environmental instrumentation. The project team first identified a wide set of possible users and services that would be interested in all or part of the VPEMS functionality. Then, the core design activities were concentrated on the development of a system able to satisfy the requirements for fleet management, including the collection of both vehicle performance (in terms of total clutch presses, brake presses, brake presses exceeding threshold, engine speed in rpm, fuel consumption, vehicle speed, traveled distance, and tire pressures) and vehicle emissions (distinguishing exhaust emissions and in-cabin emissions of total hydrocarbons, carbon monoxide, carbon dioxide, nitric oxide, oxygen, and particulate). The system was designed to be expandable to extend the VPEMS functionality in the future without the necessity of conceiving a new vehicle from the beginning [11].

There are a bundle of services, named advanced vehicle control and safety systems, that doubtless have the most ambitious goals. These goals can be enabled only if satellite navigation systems integrated with sensors and mobile communication could guarantee a very high RNP. One of them is to create a collision avoidance system through vehicle-to-vehicle communication and/or vehicle to infrastructure communications [12, 13]. Nevertheless, the mobility concept could be really revolutionized if a fully *automated highway system* (AHS) would be implemented. The concept foresees that specially equipped vehicles could travel, under fully automated control, along dedicated highway lanes without human intervention [14]. AHS has the potential to reduce fuel consumption, improve efficiency and safety, cut traveling time, and reduce congestion.

8.2.1.2 Location-Based Services

The term *location-based services* (LBS) refers to applications where the user, via a PC, PDA, or phone (mobile or fixed) sends and/or receives information that depends on the user's location [15]. Examples of indoor localization techniques are presented in Section 8.2.1.7. Position determination solutions for LBS are addressed here.

The major push factor for implementing mass-market LBS is the two directives promulgated by the U.S. FCC and the EC about emergency calls. Both directives (the European one in a softer way) oblige wireless network operators, with a certain accuracy and reliability requirement, to automatically communicate to authorities the location of a user making an emergency call. The two main technologies to satisfy the requirements are a network-based solution that relies on positioning capabilities intrinsic to the network (cell-ID, timing advance, time of arrival, or angle of arrival), and a handset-based solution that uses active participation of the handset for positioning (E-OTD, observed time difference of arrival, and assisted GPS). The second solution requires a lower investment in additional infrastructure and was largely adopted, especially in the U.S. market. This is based on GPS-enabled GSM chipsets. The main benefits of implementing LBS for service providers and network operators are:

- Location-based billing;
- Emergency caller location;
- Location-based information services.

It is expected that information services will have a fast growth over the next few years [16]. LBS will follow the trends of wireless mobile communication and could be the enabling services for 4G mobile networks. From the user point of view, it is possible to group LBS into three classes:

1. Personal navigation services;
2. Destination information;
3. Location-based advertising services.

The first class refers to all the services that help the user make decisions on movement from one place to another. It includes:

- Route guidance (indoor and outdoor);
- Public transportation information to reach a selected destination;
- Map data (indoor and outdoor);

- Friend finder services (position and route to reach another user);
- Positioning augmentation services for route guidance;
- Traffic and weather information;
- Any other information for better route choice.

The second class refers to services that give the user a better knowledge of the place in which he or she is. It includes:

- Point-of-interest information (including hotels, restaurants, and hospitals);
- Tourist information (indoor and outdoor), including information provided in museums, libraries, in front of monuments, and so on;
- Positioning augmentation services;
- Any other information service to better inform about the user site.

The third class refers to all the services that push the user to choose a new destination. It includes any type of advertising service that uses information about the location of the user.

8.2.1.3 Geographic Information System and Mapping

GIS is a computer system that stores, manages, and displays data identified according to their locations: a geographically referenced (the adopted geodetic reference system is WGS-84; see Chapter 2) database is produced with information organized into layers or themes, each one belonging to a specific topological type and relating to a specific type of data [17]. In general, a GIS can be used for environment monitoring and modeling, administration at different levels, business analysis, transport, street furniture, power and telecommunications infrastructure arrangement, emergency services, and the implementation of teaching tools. Therefore, fields of applications of GIS technology belong to all the layers (L/W, A, and S) of the model presented in this chapter's introduction; nonetheless, its main uses are recognizable in land-layer-related problems. This is the reason for discussing the topic here.

An important early use of GPS has been the provision of geophysical features, or orientation of *aerial photogrammetry*; photographs of the Earth's surface need index marks or control points on the ground to provide reference locations and to determine the scale and orientation of the photograph. Ground references or control points can be completely eliminated if the position of the camera and time of exposure are known with sufficient accuracy. This can be obtained using GPS and INSs together: this integration provides not only the exact position of the imaging device but also its orientation, so the captured images are directly related to the geodetic reference system without the need for ground control points. In practice, however, a minimum number of ground control points may be necessary to assess the resulting accuracy [18]. GPS and INS technologies can help each other reach the desired results: inertial navigation systems tend to drift over time and require reinitialization, whereas GPS has its own reference capabilities and can be an excellent augmentation for an INS, which, in turn, can help resolve cycle ambiguities in the kinematic positioning method with GPS. INS can be a valid alternative in positioning tasks during the short periods in which a sufficient number of GPS signals are not available [19].

GPS surveying methods, either kinematic GPS or real-time kinematic (RTK) GPS, offer a cost-effective solution to *cadastral problems*, overcoming related drawbacks such as extensive traversing and clear-cutting, or intervening, private properties. In unobstructed areas, the best method seems to be RTK GPS, which provides results while in the field; on the contrary, integrated solutions are suited for obstructed environments.

The use of GPS enables surveyors to deal with these problems. In addition, GPS provides user-defined coordinates in a digital format, which facilitate further analysis by means of a GIS.

Digital airborne imaging sensors and digital photogrammetric technologies allow a full development of digital photogrammetric workflow, without the need for film development and scanning. This reduces time and cost [20].

Road maps, or any other type of map, can be generated by moving over the area to be mapped: using GPS it is possible to implement a method for automatic roadway attribute entry to generate high-quality GIS databases [21]. In general, when data must be collected in digital form, the use of GPS is very advantageous: besides the mapping of roadways, other cases include digitizing the path of a pipeline or utility transmission line from helicopters, or mapping forests or streams. Beyond the use of inertial navigation systems, a video camera also can be used to address the problem of the omission of important attributes while collecting information for a GIS database. Road maps can also provide altitude profiles [i.e., *digital elevation models* (DEMs)] of the streets.

At Ohio State University, a *mobile mapping system* (MMS) has been developed to be installed in various platforms [22], for example, in a centerline painting vehicle (Figure 8.3); conventional road centerline surveys present some difficulties, such as survey crew safety, slow production rate, interruption of traffic flow, and impracticality of frequent updates of road features information. The Ohio Department of Transportation has supported the project: even if MMS accuracy would be lower than traditional survey techniques, the centerline painting vehicle used can travel at normal speed guided by the maintenance crew, avoiding the closure of

Figure 8.3   A centerline painting vehicle.

roads and enhancing personnel safety. The collected data on centerline location can be organized in a GIS database and used to supply information on location of other assets. In addition, the images acquired by the imaging module can be used to get information on pavement conditions and the possibility of repeating surveys can provide road crown data useful to trace road profiles and detect cross-sections to improve rainwater drains.

The direct georeferencing using the integrated GPS/INS can also be of use for *airborne remote sensing* and *LIght Detection And Ranging* (LIDAR), which uses an airborne laser scanner to measure the altitude of the points above the ground level [20, 23] (Figure 8.4).

Direct acquisition of DEMs is achieved by means of GPS/inertial-based position and orientation of the laser with measured altitude of the points. Furthermore, this type of LIDAR can work effectively at night, or under cloudy and high wind conditions, and can also acquire altitude measurements flying over forests, deserts, and areas covered by snow and ice [20, 24].

All-digital multisensor mapping systems that have been realized during the last few years have a fast turnaround time in producing digital image maps. The possibility of processing the acquired data in real time and mosaicking them to obtain reduced-resolution images aboard the aircraft is crucial in applications such as fighting forest fires and other emergencies. One all-digital all-direct mapping system is the DORIS [25], which integrates all of the previously mentioned technologies: a digital camera, a scanning LIDAR system, and a GPS/INS direct georeferencing system, plus a software package for acquiring and processing the data and generating image maps (Figure 8.5).

In addition, GPS/INS integration can be employed effectively to map the seafloor. The traditional way of obtaining water depth using a single-beam echo sounder installed on a survey vessel has some drawbacks: it is time-consuming and does not provide complete coverage of the seafloor. The solution to these troubles is achieved by combining a GPS/INS system with a *multibeam echo sounder*. Multiple sounding waves propagate at various angles, allowing complete coverage of the seafloor with high resolution, provided that the track lines are optimally

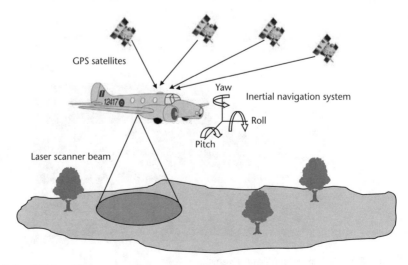

Figure 8.4 GPS for airborne mapping.

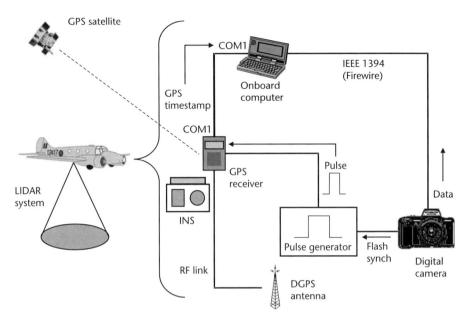

Figure 8.5 The DORIS scheme.

designed [20, 26] (Figure 8.6) and that the vessel strictly follows them. GPS navigation is used to achieve this. Multibeam echo sounders, especially those producing outer beams, require accurate positioning and attitude of the vessel and GPS/INS is used for this purpose.

Another technology that can exploit GPS/INS integration is the *airborne laser bathymetry system*. Its operation is very similar to that of a land-based airborne laser system. In LBS, the laser beam is partially reflected from the sea surface and from the seafloor, so the water depth can be calculated from the time difference between the returns of the two reflected pulses (Figure 8.7). Combining depth

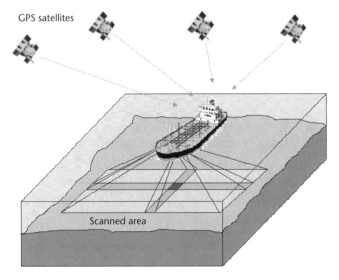

Figure 8.6 GPS for seafloor mapping.

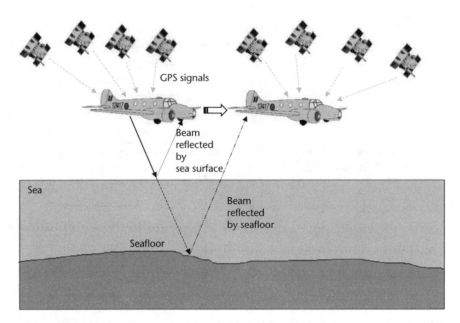

Figure 8.7 Airborne laser bathymetry system.

measurements with GPS/INS-based position and orientation of the laser, an accurate seafloor map can be obtained. The technique can also be used in difficult areas such as narrow passages. However, a drawback of this technology is that it is limited to shallow waters and is very sensitive to water clarity [20].

Moving from the seafloor and land surface to the highest layers of the atmosphere, we can also appreciate the utility of GPS in mapping the *ionosphere* [27]. The study of ionospheric behavior during both magnetically quiet and disturbed periods has traditionally been by techniques such as transionospheric radio signals, ionosonde measurements, and incoherent scattering radars. These studies are based mainly on measurements obtained either from a single station or a local network of instruments. Isolated orbiting satellites carrying ionospheric sensors have also been used. A key weakness of these approaches is the lack of instantaneous worldwide coverage.

A global monitoring system in continuous operation is required to understand the coupling processes between the magnetosphere and the ionosphere-thermosphere. Continuous operating GPS receivers can be used to great effect in this case. The refraction undergone by GPS signals can be processed to estimate the ionospheric total electron content (TEC), which is the total number of free electrons in a column with cross-sectional area of 1 square meter centered on the signal path (Figure 8.8 [27]), during quiet and disturbed periods and to investigate the energy flux dissipation and transport processes occurring in the ionosphere. The refractive index of a medium determines the velocity of propagation of an electromagnetic wave passing through it; in particular, if the medium is dispersive, as the ionosphere is, the refractive index and hence the velocity of propagation is a function of the frequency of the wave. Furthermore, in the case of ionosphere, the refractive index depends on the surrounding electron density, which varies with height as well as with latitude and longitude. After all, the velocities of the GPS signals change

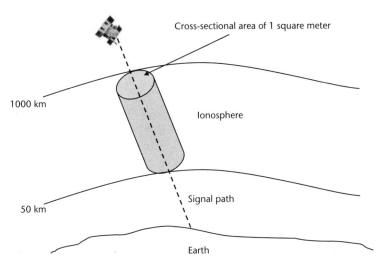

Figure 8.8   Visual definition of total electron content.

continuously as they propagate through the ionosphere. The important quantity to take into account is the total additional delay experienced by the GPS signals due to the presence of the whole ionosphere. This delay is directly proportional to the TEC, which can be obtained as already described.

GPS can also be employed to manage *distributed inventories* (e.g., steelyards where large quantities of steel are stacked in such a way to prevent warping). These stacks must be rotated periodically according to a scheduling procedure. Nevertheless, products are of different types, indistinguishable from one another, except for the position they occupy. Furthermore, physical marking is not applicable, so the location of each stack can be achieved using GPS, which retrieves data positions to be loaded on a central database (Figure 8.9).

Finally, GPS, in conjunction with GIS, is an excellent method for *wildlife tracking*. Tracking data enables the storage, in a GIS, of information related to population fluctuations or animal movement and habitat use (Figure 8.10).

8.2.1.4   Rail Transport

The use of satellite navigation in railway systems will impact all its components:

1. *Fixed assets*, such as bridges, stations, and trackside equipment;
2. *Mobile assets*, such as trains, locomotors, and carriages;
3. *Human component*, such as drivers, traffic controllers, maintenance crew, and passengers.

Some advantages of a satellite-based railway system are the following [28]:

• Maintenance expenses of the rail-track can be decreased, gradually abandoning the traditional electric and signaling outdoor equipment that is now along the track.

• Damages can be avoided, leaving out the outdoor equipment.

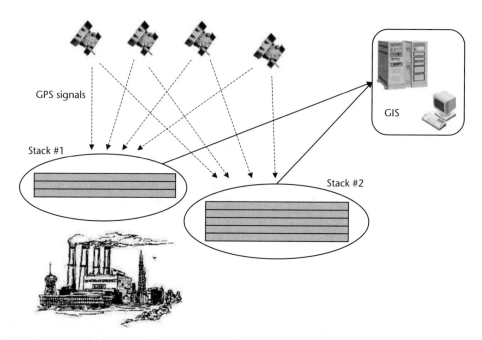

Figure 8.9   Steelyard using GPS locating capabilities.

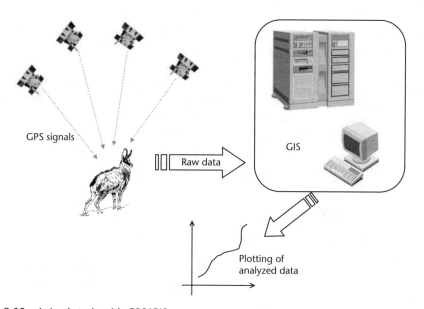

Figure 8.10   Animal study with GPS/GIS.

- Traffic safety can be enhanced on secondary lines without interlocking equipment.
- Bottlenecks can be localized, so it is possible to solve them in time.
- Transparency of the traffic flow of material and means of transport can be improved by combining and integrating different transport systems.
- Vast territories, comprising more countries, can be covered.

- Railway network can be strengthened rapidly.
- Moving vehicles can be naturally and accurately positioned and, hence, controlled.

Satellite-based railway systems can be divided into four groups [28].

The first group has *GPS-based geodesy procedures*. These were initially applied in territories requiring high accuracy but not demanding security issues. GPS-based geodesy equipment allows improvement in the accuracy of route allocation, and a comprehensive view of existing routes can be readily obtained.

The second group has *GPS-based train influence systems*, where traditional electric signaling systems are being replaced progressively by GPS-aided signaling procedures. Examples of such systems are implemented in Austria and Japan, as well as in the United States (e.g., the Union Pacific Railway Companies Positive Train Separation [29]).

The Austrian Alcatel Company has developed a new technology to control the transport on secondary lines carrying little traffic [30], hence assuring a higher level of safety than the inter-station system and reducing costs for its development. With the *secondary lines signaling system* (SLSS) (Figure 8.11) [30], the intervention of traffic controllers in the management of tracks is reduced and the system can store information about its own status automatically. The SLSS has different functions: the easiest one is the location of the train position and its display on the monitor installed in the control center; the other one is the GPS-based train command center. The traffic controller can give dispositions to the trains or the whole line can be managed by a timetable-based computer. Alcatel uses a train radio to exchange data between the control center computer and onboard equipment. Finally, the traditional signals along the track, indicating speed limits and braking points, can be eliminated, given that a monitor in the driver cab displays this information.

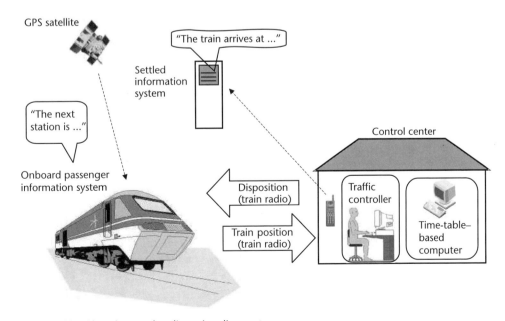

Figure 8.11   Alcatel secondary lines signaling system.

The Japanese railway also has a GPS-based security system, called CARAT (Computer And Radio Aided Train) [31], whose main function is to protect the maintenance group on the tracks. Each group leader has warning equipment that alerts workers about the train's arrival, so they can leave the track safely beforehand. Another advantage of the system is represented by the lower costs due to better driver (engine) performance (e.g., avoiding sudden reduction and increase in speed).

In the third group are *railway transport systems*. In Germany, the *Satelliten-Handy* (Sandy) [32] has been realized with the basic idea of preventing the damages of the goods, which are caused not only by stealing but also mostly by the shunting crew. The Sandy is thief-proof and can be mounted on the carriage to collect parameters continuously and transfer them to the system operator via SMS or by train radio at preestablished time intervals. When a measured value of the controlled parameters is higher than a decided threshold value, Sandy will send an alarm message to the operator, so as to evoke the necessary steps to solve the problem.

The fourth group deals with the *GPS-based passenger information systems*, which, as in the case of the *S-Bahn Berlin system* (Figure 8.12), can be supplemented with a communication system [33]. The passengers use it only in emergency situations, going to the settled passenger information and supplication box or asking for help on the board of the S-Bahn train by pressing a button. Then the security service can be guided from the command center to the GPS-determined place.

Talking about the *onboard information equipment*, a satellite-based central system can inform passengers of the name of the next station, the remaining time to the next stop, and the possible late arrival using TV screens or LCD monitors. This information can be supplied at a desired frequency. In addition, information about sights along the track can be given by using interactive monitors (Figure 8.13), which also show carriers controlling their own transport [28].

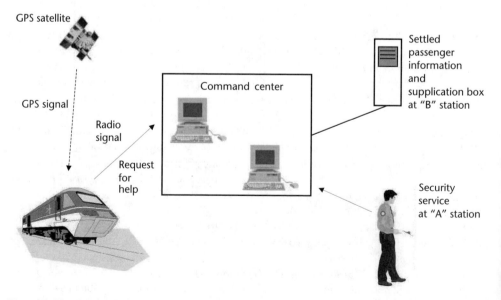

Figure 8.12   S-Bahn Berlin system overview.

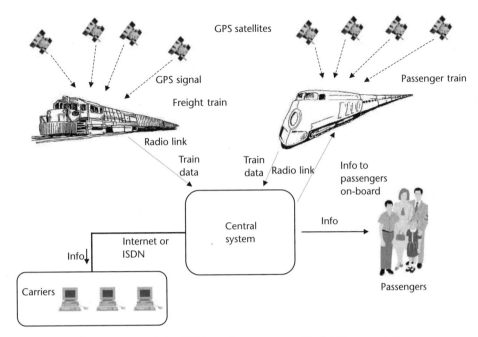

Figure 8.13 Informatogram of the GPS-based passenger and freight transportation.

Another important application of the satellite-based passenger systems is *time-table planning*: storing the time differences between the timetable and the real data, it is possible to create a database for timetable planning, making the work of the planning engineers easier and more accurate.

Also the railway *freight transportation* can be supported by GPS, according to two different ways of identifying carriages [28]. We refer to *one group train*, when all carriages must arrive at the same destination, hence representing a single element from a transportation point of view. In this case, tracking the position of the engine is sufficient to know the position of one of the carriages with an accuracy equal to the length of the whole train (including the engine and all carriages), whereas the exact positioning data of each carriage is not important for the carriers. There is a weakness in this system. To connect the carriages logically to the GPS-based engine, it is necessary to know their identification numbers; before the departure, the numbers have to be collected and transmitted by radio to the system. When the train stops and when a carriage is uncoupled several times and later sent on with another train, the risk of missing data increases.

The other option is summarized by the expression *carriages following concept*, in which every wagon has its own GPS-based identification equipment, hence avoiding the problem of missing data during the "radio-guided carriage data collection." The mobile equipment sends SMS messages to the engine, thereby reducing costs of transmission.

In both previous cases, the engine will transmit train data to the traffic controller by means of train radio. Knowing the difference between actual time and timetable data, the traffic controller can be informed and can detect possible late arrival. This information could then be used both to reduce lateness and to inform commuters of the revised time of arrival (Figure 8.14).

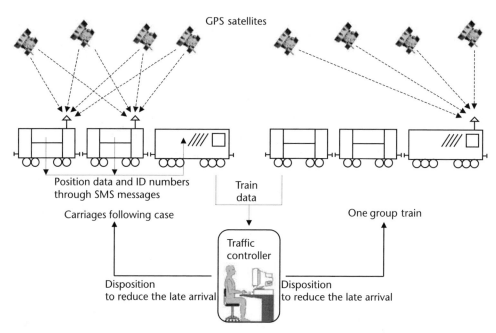

Figure 8.14 Rail freight transportation types.

Furthermore, using GPS, carriers can follow their freight in real time, whereas, with earlier systems, they could only know whether a wagon was under transport or not.

In Europe, the *European Rail Traffic Management System* (ERTMS) is becoming the standard for train control, signaling, and traffic management [17].

Satellite navigation has an important role in two layers of ERTMS, the *European Train Control System* (ETCS), which deals with train control and protection, and the *European Traffic Management Layer* (ETML), which deals with nonsafety-of-life-related aspects like traffic management and regulation.

The introduction of satellite navigation within the ETCS/ERTMS will mainly help to improve performance on high-density lines and reduce costs on low-density and regional lines [34].

GALILEO, which is a civilian system, provides characteristics for certification, operational transparency, and service guarantee that give it a primary role in rail transport. The real-time navigation performance integrity monitoring (*integrity flags*) over the service area makes the GALILEO system suitable for all *safety-of-life* requirements.

Several projects and studies have been promoted by the EC and the ESA with the aim of developing GNSS safety-related applications (Table 8.2) [35].

The high safety requirement can be satisfied by a hybridization of GALILEO receivers with other sensors, such as odometers, balises, and gyroscopes.

In the *nonsafety-of-life* application sector, many national train fleets use satellite navigation for fleet management, including certain U.K. fleets (mainly freight), SNCF in France, and DB in Germany.

Ferrocarriles Generalitat Catalunya (FGC) in Spain is realizing the *SITCAS system*, which allows the control of train location and speed in real time.

Table 8.2 Summary of EC/ESA GNSS Safety-Related Projects for Rail Transport

Project	Subject	Trials Test Date
GADEROS	GNSS for train control system according to ERTMS standards	Second half of 2003
INTEGRAIL	EGNOS for traffic management and control	Different demonstrations in April–June 2003
ECORAIL	Train positioning for traffic management	September 2003
LOCOLOC	Low-cost train location system for low-density lines	Mid-2004
RUNE	Enhancement of train positioning and speed sensors through GNSS complement	June 2003

Kaiser-Threde in Germany uses existing satellite navigation infrastructures in the *Railtrac-Kt system* for the supervision and localization of wagons, containers, construction machines, and other mobile objects. GSM or satellite communications are used to report unplanned stops or faults.

In the *safety-of-life* sector, researchers are developing applications for proximity alarms, which warn when two trains on the same track are too close, and velocity limit alarms, which monitor violations of speed limits for a section of the track [17].

The role of GALILEO can also be important from an *energy optimization* point of view. Drivers often change speed without concern for saving energy, for example, braking sharply before a tunnel instead of using regenerative braking at the appropriate distance before the tunnel. The knowledge of train position with respect to its environment is fundamental and can be obtained cheaply from satellites [34]. A better-planned breaking process improves the traction parameters, resulting in less traction energy consumption and reduced noise and dust pollution.

Surveying track status is an important task for ensuring safe passage for trains. A good survey needs accurate position determination and synchronization between the positioning system and other testing/inspection systems [36].

More sophisticated applications are being developed, for example, to improve *passenger comfort,* through the study of the train tilting in curves, combining data from accelerometers and other sensors with information coming from satellites [17].

Train simulators have been employed for training drivers to enhance their driving skills and their ability to reduce fuel consumption. The simulators need data, such as train composition and loading, locomotive capacity, video imagery of the track under simulation, and geometrical data defining the track. All this information can be rapidly acquired through the use of GPS, which can provide the three-dimensional location of the train with the desired precision. Its use requires continuous reception of signals from a minimum of four satellites, a major limitation in the circumstances of rapid track mapping. Data provided by GPS can be integrated with inertial instruments, such as grade sensors, spinning mass gyros, and pulse generators mounted on an axle [37], to remedy the possible lack of sufficient number and/or power of GPS signals.

The *Alaska Railroad Corporation* (ARRC) recently implemented the *collision avoidance system* (CAS), based on the integration between real-time GPS positioning and VHF data radio communications. The aim is to prevent train-to-train collisions, protect track maintenance forces, and enforce speed limits [38]. Managing train speed and enforcing train movement regulation (e.g., the right to occupy and operate within a section of the track) are crucial elements of the system. The onboard system continuously checks the vehicle location, comparing it to the track it is authorized to occupy. Braking algorithms are applied whenever the train reaches speed values greater than those allowed for the track segment it is occupying.

8.2.1.5   Pedestrian Navigation

Pedestrian navigation includes all services and applications where the user is a pedestrian or a mobile unit at very low speed, equipped with a portable receiver. In this respect, this field of applications is characterized by specific requirements on the size, weight, and ergonomy of the navigation device. In addition, since this type of user can either stand still or move in both open spaces and indoor environment, pedestrian navigation has to provide accurate positioning performance in both outdoor and indoor situations. Considering the other navigation application scenarios, these structural, environmental, and performance features lead to the necessity to develop a new specific approach, which includes the capability to provide navigation services both in the presence and absence of a satellite navigation signal. In fact, since the typical application scenario for pedestrian navigation is the urban environment, the probability of working without the satellite navigation signal is high. This is mainly due to the shadowing effect, caused by buildings or other impairments around the pedestrian user receiver and, in particular, signal attenuation/blockage indoors. The pedestrian user, hence, typically operates in signal-blocked environments, which result in a low availability of the satellite navigation signal. To overcome this difficulty, current studies are focused on portable devices that use satellite navigation systems as main positioning modules, but, in addition, they provide positioning through alternative navigation techniques, essentially based on DR calculations [39, 40]. The introduction of these techniques has a dual aim. First, in case of availability of satellite navigation signals, they improve positioning accuracy, as discussed in [39], where it is shown that, in the absence of magnetic field disturbance, a magnetic compass with bias correction from a Kalman filter integration process of GPS data and DR sensors can provide the correct heading, leading to a higher level of positioning accuracy. Second, in case of signal absence, they back up the satellite navigation systems, ensuring the service to the users. Pedestrian navigation finds its main application in the social context, where GNSSs are used to assist impaired people. Three social applications currently under study and development are highlighted here: *blind persons, Alzheimer's sufferers,* and *people with motor impairments.*

People with impaired vision depend largely on the additional aid of sighted persons, since the handicap limits their ability to autonomously move in unknown and dangerous environments, such as urban centers.

Current studies by American and European researchers, respectively through GPS and EGNOS/GALILEO technologies, are focused on this problem, evaluating the possibility of using satellite navigation technologies to help blind people [41–44].

ESA is involved in an innovative ongoing project that aims at improving the GPS-based personal navigator for people with impaired vision. Called *TORMES* (Figure 8.15), the project is being developed by GMV Sistemas in Valladolid, Spain, and the Spanish national organization for the blind ONCE (Organización Nacional de Ciegos Espaòoles) [41].

The TORMES tool will be enhanced with the new European satellite navigation technologies, EGNOS and SISNET (see Chapter 4), which significantly improve its performance. In particular EGNOS corrections improve GPS accuracy up to 2m and SISNET technology enhances the EGNOS service, providing EGNOS data through the Internet via a GSM connection. The new handheld device, hence, will be able to provide users with their positions, routing, and guidance, ensuring high performance in every condition, as a funny and ambitious declaration by Jose Luis Fernandez Coya, the head of ONCE's research and development department, said: "When blind people take a taxi, they will be able to give directions to the taxi driver!"

Navigation satellite technologies can be applied to provide assistance to Alzheimer's sufferers. This application is highlighted next, distinguishing between patients in the early and advanced stage of their illness.

In the first case, Alzheimer's sufferers are able to live a near-normal life, although they suffer sudden memory losses. A portable satellite-based navigation tool (PDA) programmed with personal user information can provide assistance to the sufferers, allowing them to autonomously determine their position and, with the aim of a simple touch-screen interface, find the right way [44].

In the second case, taking into account patients with more severe symptoms, satellite technology can be used to monitor Alzheimer's sufferers to provide assistance to them in case of need.

Figure 8.15 The TORMES tool.

Nowadays, possibilities of free and autonomous movement of disabled people is largely limited because of either the lack of suitable structures or, in case of their presence, the difficulty in identifying and reaching them easily and on time.

For the latter feature, ongoing studies are evaluating the possibility of adapting the satellite navigation tool currently used in the vehicular environment for providing route planning for people with motor impairments [44]. The system, working in cooperation with a dedicated GIS for disabled people, will provide useful information to the user, such as specific points of interest, special parking places, positioning of dedicated structures, and many other guidance and routing services.

8.2.1.6 Leisure

Among the several application scenarios for positioning systems, the leisure segment has been gaining a role of increasing importance in the navigation market during the last few years. This sector, in fact, is expected to see a remarkable development of applications that can be only partially imagined at the moment. In particular, the new capabilities provided by GALILEO and the modernized GPS (see Chapters 5 and 6) will allow the development of new, unexpected, and unusual classes of services and applications, which will be added to the current recreational applications, such as hiking, orienting, flying, and sailing. The following highlights some of these new and particular application challenges; the authors consider it useful for readers to study these innovative scenarios rather than those already widely tackled in several publications.

The world of sport has always been supportive of new technologies since the capability to provide spectators with timely and detailed information about a specific race or match significantly contributes to the improvement of spectator (viewer) enjoyment and worldwide take-up of sport.

For this reason, some companies are developing systems that aim to enhance the sporting events, employing the capabilities of precise positioning, velocity, and timing provided by satellite navigation systems.

In this respect, the New York company *Sportvision*, which specializes in the television sporting sector, has developed, in cooperation with *NovAtel Inc.*, a GPS-based system, named *RaceFX*, that allows tracking and displaying on screen the real-time location of all cars throughout a racing event [45]. The new tool was introduced for the first time in 2001 during a Daytona 500 event.

Specifically, during a racing event, the system can increase racing fans' enjoyment by providing real-time graphics that describe the position of the race cars, their speed, and their time parameters. Furthermore, the system provides other interesting statistics, such as the path of the fastest car, that significantly improve viewer understanding of the event.

The RaceFX system architecture is displayed in Figure 8.16. The RaceFX system consists of four subsystems: GPS, telemetry, time synchronization, and video overlay. The first subsystem is the system's core, since it allows cars to autonomously and constantly determine their position during the racing event. The race cars, in fact, are equipped with a GPS receiver and a transmitter/receiver device at 900 MHz, which transmits car data to the control center located within the in situ television broadcast center, through a 900-MHz communications network located around the

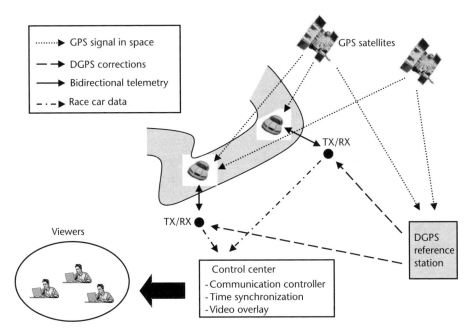

Figure 8.16 RaceFX system architecture.

racing track. The control center processes the received data and generates the statistics and graphics that will be displayed on the viewer screen.

The Sportvision system also includes a sophisticated telemetry system that provides a bidirectional link. In downlink, the telemetry transmits DGPS correction data from a GPS reference station located near the track to the cars at 0.5 Hz, whereas, in uplink, it conveys car data information to the video subsystem at 5 Hz. RaceFX, in fact, suffers from constraints and problems typical of satellite navigation-based systems, such as the need for contemporary visibility of at least four satellites (see Chapter 2) for providing three-dimensional positioning and DOP features. Considering the peculiarity of the racing track scenario, these topics are considerably critical aspects for this application. The use of the DPGS technique aims at overcoming these issues to provide the required accuracy. In addition, especially referring to the shadowing problem due to the signal-blocked environment, NovAtel and Sportvision engineers have developed several techniques based on a computerized model of the track to reduce its effects and, hence, improve system performance.

An analog GPS-based technology was applied to *ski races* [46]. The tool, developed by a group of Swiss engineers and tested in January 2001 at Lauberhorn, Switzerland, during the Wengen World Cup downhill course, can improve training procedures by determining the best line and checking the skier performance over the ski slope. Figure 8.17 shows a device which consists of a waist harness and a helmet containing, respectively, the GPS receiver and antenna.

Another interesting tool was developed by Finnish company *Suunto* that designed a sport watch to help golfers during their match or training by providing them with useful information about clubs, games, and the course [47]. The watch, named *G9*, uses GPS technology for time synchronization.

Figure 8.17   The GPS-based Swiss tool for skiers (2001).

These devices are only some examples of the possible sport applications of satellite navigation, but they show how the range of applications can be wide in this environment. The introduction of these technologies, in fact, aims at giving athletes, their coaches, and television viewers a new analysis tool to enhance and improve performance (of both athletes and the viewer experience) [48].

The tourist segment represents another sector in which the capabilities provided by satellite navigation systems can be largely employed [49–51]. In fact, the knowledge of their position allows tourists to optimize their time by permitting them to move in an unknown place and reach tourist destinations more easily.

Another application challenge is offered by self-guided tours. Visitors are equipped with a portable device, which includes the satellite navigation receiver and a memory with information about the visited place. When visitors are next to a tour point of interest, the device automatically starts to give the relevant information. Therefore, the use of satellite navigation technology allows the conduction of such tours without the need for any code of reference for the given location, as currently required. The tool can also provide users with other useful positioning information to improve the tour's quality. Nowadays, this kind of technology, based on GPS, is offered in some locales, such as Yellowstone Park in the United States.

8.2.1.7   Indoor Applications

Indoor environments are challenging for navigation applications. Hence, it is necessary to implement customized technical solutions, combining navigation systems with mobile communications and data transmission systems. The applicability of satellite navigation techniques in indoor environments is highly thwarted by attenuation effects on signals or even complete obstruction by walls, low SNR, and strong multipath effects.

The candidate *systems for indoor positioning* can be divided into three groups [35]:

1. *Satellite navigation* (GPS, GLONASS, GALILEO) stand-alone or integrated with augmentation systems (EGNOS, WAAS, MSAS) and/or satellite-navigation-like techniques (for example, pseudolites);
2. *Mobile communication and radio systems* (GSM, EDGE, GPRS, UMTS, WCDMA);

3. *Short-range radio systems* (Bluetooth, IrDA, WLAN, DECT).

Stand-alone satellite navigation in indoor environments can be effective if one or more of the following objectives are reached:

1. A greater number of available signal sources, for example, by means of installation of pseudolites (pseudosatellites), which are GPS-like signal generators, so as to allow pseudolite-compatible receivers to be realized with minimal modifications to existing GPS receivers;
2. Utilization of signals and/or carriers that better propagate and penetrate in indoor environments;
3. Adoption of new tracking and demodulation techniques;
4. Introduction of network-assisted techniques to reduce the number of necessary satellite signals to perform the positioning task;
5. Reduction of tracking times by using pilot tones.

With regard to the installation of *pseudolites*, it has been demonstrated this can enable an accuracy of few centimeters to be achieved [52]. In 1999 the Seoul National University GPS Lab (SNUGL) started the implementation of a miniature vehicle equipped with two GPS antennas, which would be able to move autonomously in an indoor environment. The whole system (Figure 8.18) also included a reference station and pseudolites, which were fixed in positions in the ceiling determined offline [52].

The reference station was fixed on the floor and used to transmit, by means of a wireless datalink, carrier phase corrections to the vehicle, which was able to calculate its position using carrier phase differential GPS. Unlike outdoor environments, indoor navigation must face the near/far problem, according to which the received power from closer transmitters interferes with those further away. Furthermore, indoor signal propagation is affected by strong multipath. Besides these general problems, in the case of pseudolites, the question of time synchronization also arises (clocks used are temperature-compensated crystal oscillators, or TCXO). At SNUGL, a "master" pseudolite had been installed at the center of the ceiling to synchronize the receiver's sampling time by its navigation messages. The solution of the

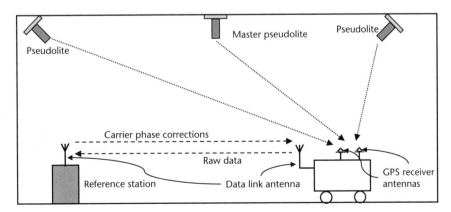

Figure 8.18 Overview of the system developed at SNUGL.

near/far problem had been obtained by tuning the pseudolite signal power and adopting a pulsing scheme [53]. Multipath was mitigated by controlling the gain patterns of the pseudolite antennas. Finally, a cycle-slip recovery and automatic cycle-ambiguity resolution functions were implemented.

However, the use of pseudolites has difficulties both in complex buildings, such as office buildings, and in crowded buildings, such as shopping centers. The problem in both cases is the attenuation of signals. Hence, pseudolites offer high-precision navigation in particular environments and indoors in single large rooms (e.g., production halls, hangars, large factories, indoor amusement parks, or for indoor E-911).

In general, user requests for guidance in an indoor environment refer to a specific target more than to a point of coordinates. Therefore, an *indoor guidance and information system* (IGIS) has to supply the user with the determination of his/her position and navigation and guidance to a desired destination on an optimal path. Furthermore, an IGIS has to be able to provide information about mobility support, tourist information, e-payment and e-commerce, event-specific real-time information, and so on.

An IGIS has to work in conjunction with outdoor navigation systems, providing a seamless transition capability between outdoor and indoor environments. A generic architecture of an IGIS consists of elements pertaining to different transmission technologies, such as mobile communication, short-range radio and satellite navigation, plus an IGIS provider, which attends to data collection and data traffic management, and the user terminal as a graphical user interface to the system (Figure 8.19).

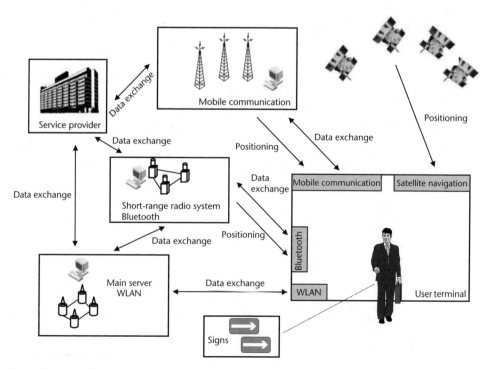

Figure 8.19   System architecture of an IGIS.

As shown, satellite navigation is just one technique for positioning and guidance in an indoor environment and has to be used together with other techniques and methods.

An effective alternative to pseudolites is represented by the method of *reradiation of GPS signals* [54]. These are received on the ground surface to be demodulated and divided into individual satellite signals, which are then modulated separately to be transmitted, for example, through optical fibers or coaxial cables, to reradiation antennas set at appropriate positions in a target space, such as an underground (Figure 8.20).

The indoor GPS receiver measures a range that is the sum of three components:

1. The range between the satellite and the surface GPS receiver;
2. The length of the cable that connects the surface GPS receiver and the reradiation antenna;
3. The distance between the reradiation antenna and the indoor GPS receiver.

Since the positions of the reradiation antennas are known, the position of the indoor GPS receiver can be calculated. A reference receiver can also be introduced in the indoor environment with the GPS receiver, whose position must be determined. Applying a relative measurement, the position of the GPS receiver can be obtained as if the satellites were at the point of the reradiation antennas [54].

In indoor environments, as already outlined, the problems to solve are:

• *Near/far problem:* This one can be worked out by adopting pulsing of GPS signals, changing the radiation pattern of the reradiation antennas, or increasing the dynamic range of the receiver.

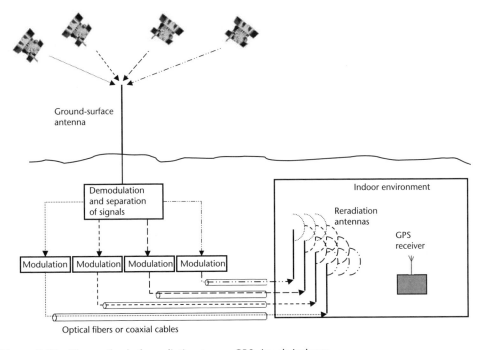

Figure 8.20   The method of reradiation to use GPS signals indoors.

- *Multireflection:* Determining the phase-range rather than the pseudorange can lead to better results.
- *Antenna interference:* If the reradiation antennas are distant from each other, the interference should be weak; if they are placed close, CDMA can be used; the interference between ground surface antennas may not give practical problems.

Indoor-capability of GPS can be achieved by means of *assisted-GPS* (A-GPS) in combination with *massive parallel correlation.*

Standard GPS receivers take one to two minutes to search for and acquire satellites, but they cannot lock on them if the signals are attenuated, as occurs in indoor environments. In fact, when performing a cold-start, a standard GPS receiver must do a search both in frequency and in code space with only two correlators (one early and one late) per channel (i.e., per satellite), with which they can search sequentially the 1,023 possible code delay chips, in each adjacent frequency bin (Figure 8.21). The dwell time in each frequency/code bin must be no longer than 1 ms to not take too much time to complete the whole search. However, a short dwell period limits the detectable signal strength.

A-GPS had been introduced to point the receiver to the frequency bins to search (Figure 8.22), hence helping the receiver to either shorten the acquisition time or dwell longer in each frequency/code bin, reaching higher sensitivity but not sufficient enough to detect the weak GPS signals in an indoor environment.

To achieve the processing gain necessary to detect satisfactorily GPS signals indoors, adopting massive parallel correlation is required. Searching all possible code delays in parallel, the receiver can dwell for much longer in each possible frequency/code bin, resulting in more detectable signal strength [55].

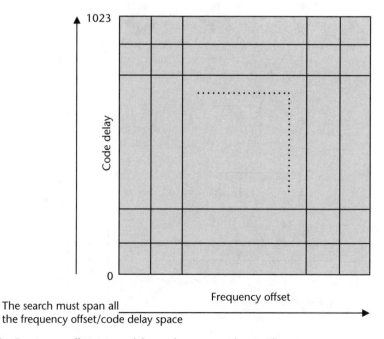

Figure 8.21   Frequency offset versus delay code space, without aiding.

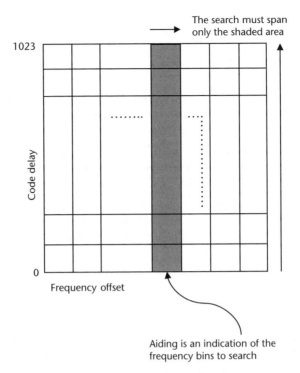

The search must span
only the shaded area

1023

Code delay

0

Frequency offset

Aiding is an indication of the
frequency bins to search

Figure 8.22   Frequency offset versus delay code space, with aiding.

However, reducing frequency search space attained by A-GPS means that the broadcast orbit information is valid for only 2 to 4 hours, and any other information, such as the expected Doppler shift, is derived only from the data transmitted by satellites. An implementation of both A-GPS and massive parallel correlation was created by Global Locate, which put into practice a worldwide reference network to supply assistance data valid for 10 days [55]. This network tracks all the GPS satellites all the time; Global Locate can synthesize orbit models, which are able to provide A-GPS assistance data with a usability time span 80 times greater than broadcast ephemeris.

The importance of parallel correlation for performance is clear. The application of indoor GPS in cell phones raises the concern of the effect of a greater number of correlators on the phones' CPU. In traditional GPS receivers, the CPU has to read, store, and process correlation results in real time and its load increases with the number of correlators; also the integration with other software tasks is difficult due to the interrupt-driven processing that must run at high priority. After all, in traditional GPS receiver architecture it is very hard to host successfully GPS software in a wireless CPU. The solution proposed by Global Locate reduces the software role in signal tracking and processing, giving more functionality to dedicated hardware (Figure 8.23) [56], which sends the results of convolution and integration to the host CPU, where a nonreal-time driver determines GPS pseudoranges.

The characteristic of cell phones' portability also suggests the problem of power consumption. The parallel operation of a lot of correlators pulls down the amount of necessary power, the time employed in correlation being shorter than that needed

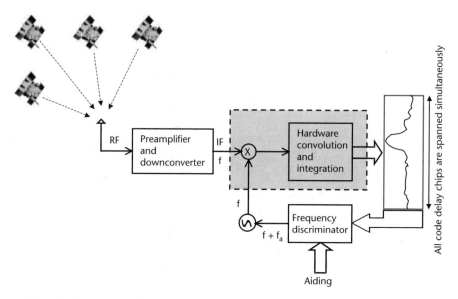

Figure 8.23 Hardware processing approach for indoor GPS with aiding and massive parallel convolution.

by a smaller number of correlators operating in sequence, with energy consumed being the same.

Indoor GPS performance can receive a helping hand by improving the *receiver frequency reference*, which is, usually, a quartz crystal oscillator [57]. Mass-produced cell phone TCXOs can present anomalies, such as microjumps and activity dips, which are both changes in frequency, with the difference that the latter ones always take place at approximately the same temperature, given a particular crystal design, but are much slower. Microjumps become critical when they occur in safety-related applications, such as E911. Therefore, optimal receiver performance can be achieved by removing these frequency drifts and drift rates from the search process to account for the increased criticality of the indoor environment.

8.2.1.8 Precision Agriculture

In the last few years, the importance of precision agriculture has been increasing in the research and manufacturing world. Continuing improvements in satellite navigation systems, particularly in terms of accuracy performance, have opened new perspectives and opportunities for using these technologies in this area (see Chapter 4). Furthermore, these improvements have made the development of accurate GNSS-based automatic guidance systems possible, with the potential of reducing human workload [58].

GNSS applications in precision agriculture include *chemical spray control*, *soil health monitoring*, *crop yield monitoring*, *crop acreage*, and *livestock tracking* [20, 59, 60].

Concerning the first application, precise management of chemical spraying can be obtained by equipping specific aircraft with a GNSS-based aerial guidance system. Precise positioning facilitates optimal spraying and improves the chemical spraying process significantly. This allows pilots to spray fertilizers, insecticides, or herbicides only on the wanted area, minimizing, hence, possible human errors or overlaps and, at the same time, optimizing use of fuel and chemical products. Therefore, this technique contributes to safeguarding the environment and allows better management of agricultural land.

Soil health control is another key aspect for commercial farms since soil quality is important to production efficiency. In this respect, GNSS capabilities can be employed to accurately locate the place of origin of a soil sample to have a precise and detailed outline of the status of agricultural parcel. Furthermore, this information can be used to develop reference guides aiming at helping farmers treat soil problems in an efficient and cost-effective manner.

GNSS technology can also be used to monitor crop yields. The technique enables one to develop crop yield maps by equipping harvesters with a GNSS-based tool, which combines current yield rates with position information. The maps should indicate the specific level of productivity for each crop acreage, enabling farmers to take immediate action in case of an unsatisfactory level of yield.

The last application discussed uses satellite navigation systems to track crop acreage and livestock. Replacing the traditional measurement technique based on tapes and wheels, satellite navigation system capabilities can be used to provide farmers with information about the extension of their parcels and crop acreage. In addition, by using transponders linked to a central database, the introduction of satellite navigation systems enables farmers to continuously trace livestock and products. This makes work easier and faster for farmers and allows them to improve the management and control of their property.

In light of these considerations, introducing GNSS-based technologies in the agricultural sector aims at increasing productivity, evaluated in terms of quality, quantity, and cost-effectiveness of farm produce. This results in several benefits for both manufacturers and customers.

8.2.1.9 Energy

An interesting application of GNSS, identified among the ones envisaged for GALILEO and modernized GPS systems, is related to electricity [61, 62]. In particular, the very precise timing data obtainable via GALILEO will help optimize the transfer of electricity along power lines. GALILEO could also help in the maintenance of the electricity-distribution infrastructure. Power grids are continuously monitored by a range of instruments spread around the system. Information from these instruments is used to repair the system when a power line breaks or weaknesses appear in the grid. GALILEO will improve the instrument time synchronization to provide a more rapid return to full service [61].

8.2.1.10 Civil Engineering and Other Applications

Civil engineering is another environment where the use of GNSS capabilities is significant. In this area the main applications concern *precise structural placement, structural displacement* and *deformation monitoring, road construction, Earth moving,* and *fleet management* [20, 63, 64]. Specifically, the combined use of GNSS technology and other communications and computer systems can improve quality, cost-effectiveness, and speed of the operators in performing their work.

Further interesting exploitation of GNSS is provided by the following application examples:

- *Weather forecasts* by using the meteorological data stored in the GPS standard RINEX (see Chapter 7), as developed in the GPS Atmosphere Sounding Project (GASP), which uses GPS to continuously monitor water vapor on both regional and global bases [65];
- *High-precision crane guidance* in container terminals by means of GPS-based auto-steering system [66];
- *Automatic vehicle location in the World Trade Center recovery setting* to follow the trucks employed in debris removal along the whole journey, from loading materials at the site to unloading at disposal sites [67];
- *Air pollution monitoring* using city buses supplied with air quality sensors and GPS receivers to collect air pollutant data, associating them with a time stamp and the current bus position—the aim is to implement a proactive traffic management system to redirect traffic flow to lower-polluted sectors [68];
- *Kinematic survey of archaeological sites* to determine the best archaeological strategy, avoiding useless excavations [69];
- *Location-based encryption/decryption* to provide a further level of security beyond that of conventional cryptography: position and time are used to cipher a plain text in such a way that the cipher text can be decrypted only at a specified location [70].

8.2.2 Water

With the aim of differentiating and quantifying marine navigation safety requirements in the United States, the Transportation and Defense Departments have defined marine navigation in terms of phases. The four phases defined in the Federal Radionavigation Plan are *ocean, coastal, harbor/harbor approach* (HHA), and *inland water:* Each phase is characterized by a remarkably different set of performance requirements based on safety and environmental concerns and aimed at minimizing marine collisions, ramming, and grounding.

GPS is able to satisfy many of the ocean and coastal phase performance requirements; HHA and inland waterway navigation have different and somehow tighter performance requirements that rendered GPS unacceptable until the removal of the selective availability in 2000. The main distinction in terms of performance requirements among the maritime navigation phases is accuracy; in particular 1,800–3,700m is the accuracy required for the ocean phase, 460m for the coastal phase, down to 8–20m for HHA, and even better for inland waters.

In the HHA phase, a vessel pilot requires accurate, frequent, and timely verification of the position. Deviation from the desired vessel track drives the pilot's decision-making process. This is quite different from the ocean phase, where position updates in the order of minutes are satisfactory. The need for frequent position verification results in additional burden on the radio-navigation service provider.

Sudden and unpredictable signal outages can significantly jeopardize the safety of vessels involved in the execution of sensitive maneuvers. An availability specification of 0.997 is given for HHA. Nonetheless, these specifications must be used with some caution. In fact, the stated accuracy and update specification lack specificity with respect to dynamic conditions under which a stated level of sensor performance shall be provided. Furthermore, no explicit allowance is provided for pilot error contributed by personnel skill, pathway geometry, ship dynamics, and disturbances. The concept of "technical error," envisaged in aircraft guidance, is not systematically exploited in marine navigation design.

Radio-navigation requirements depend on vessel size, maneuvering activity, and geographic constraints of the operating area. Ship pilotage may be unsafe, even with perfect navigation sensors, if maneuvers are complicated by tight channel tolerances, unexpected traffic, poor helm dynamics, and strong wind and current disturbances.

In exploiting GNSS—currently GPS and DGPS—for the marine environment, particular attention has to be given to vessel footprint steering performance and the interplay between sensor and ship models [71–75]. Other related functions, such as hazard warning, risk assessment, and on-line dynamics modeling, are also important elements.

It is worth mentioning here some background on the experimentation that led to the deployment of a standardized marine DGPS service in the United States [76]. In the early 1980s, the DOT began studying the potential civil use of GPS, quickly realizing that many potential applications would require higher levels of accuracy and integrity than SPS would be able to provide [77]. The feasibility of providing DGPS for marine navigation with an absolute accuracy of 10m was demonstrated in 1987. In 1989 the Montauk Point (New York) radio beacon was temporarily converted to provide a DGPS test broadcast. Given the success of these 1989 tests and the desire to evaluate DGPS in operational environments more thoroughly, a prototype DGPS service was established in 1990. The Montauk Point marine radio beacon then became part of the U.S. Coast Guard Northeast test-bed prototype service, providing accurate navigation from Cape Hatteras (North Carolina) to Canada; after that, the USCG covered most of the coastal areas in the United States.

An essential part of the coastal service is a radio link for transmitting differential corrections. An approach that has been extensively adopted is the minimum shift keying (MSK) digital transmission of data over the existing network of low-frequency marine radio beacons. The radio-modulation format was selected to minimize interference with the radio direction finding function currently provided by these beacons. In addition, forward error detection and correction features have been employed to reduce the impact of noise, resulting in a highly reliable link, even beyond the normal range of radio beacon service [71]. In 1987, an assessment of DGPS, differential LORAN-C, and other candidates for the Great Lakes and St. Lawrence Seaway was carried out under U.S. Maritime Administration sponsorship. Furthermore, the interplay between sensor dynamics/noise characteristics and ship control performance in an automated steering environment has been explored. The steering characteristics of DGPS and other radio-navigation sensor combinations were integrated with an augmented state navigation filter, and an optimal steering controller was developed [76]. In 1988 the Coast Guard Research &

Development Center (RDC) conducted human factor simulations with experienced pilots operating with simulated radio-navigation sensors to validate the Federal Radionavigation Plan requirement for 8–20-m accuracy for harbor and harbor approach. As a conclusion, also reported in other investigations the use of DGPS for steering in confined waterways resulted in the need for careful integration of risk assessment and hazard warning functions collecting information from other sources such as radar, geographic databases, and shore side vessel traffic systems (VTS).

The issue also concerns *underwater* navigation, as many possible autonomous underwater vehicle (AUV) missions require a high degree of navigational accuracy. GPS is capable of providing this accuracy but it has to keep into account that intermittent reception caused by either wave action or deliberate submergence causes the loss of GPS position fix information for periods extending from several seconds to minutes. A system has been designed to show the feasibility of using a low-cost strapped-down inertial measurement unit (IMU) to navigate between GPS fixes [78].

As outlined in Chapter 5, maritime navigation is also one of the envisaged applications of GALILEO. In fact, open ocean and inland waterways are the most widely used modes for transporting goods worldwide and goods transportation is one of the strategic aims of the GALILEO-based transport network. Furthermore, the high accuracy and integrity, certified services, and high availability brought by GALILEO will be applied to leisure boats, commercial vessels, and all ships falling under the *safety-of-life at sea* (SOLAS) *convention* in every phase of marine navigation and in all weather conditions [17].

In addition to these applications, systems are under development to equip containers with standardized positioning devices that allow better logistics operations; in fact, handling of containers is crucial for efficient commercial harbor operations and for reaching and managing effectively approximately 40 million containers per year in European ports.

8.2.2.1 Fishing

In recent years, the needs of the fishing sector have grown, ranging from day-by-day operational support to the navigation and positioning of fishing vessels. Strict international rules governing the intrusion into national waters demand that vessels are monitored to check they work only in designated areas.

The long journeys and global nature of fishing activities mean that satellites provide the only viable and reliable method of navigation; the continuous views of the sky also renders high seas an ideal environment for exploiting a satellite-based navigation means. Modern fishing vessels (Figure 8.24) travel the globe pursuing commercial returns, at the same time sending regular position reports (position, speed, heading) and occasional information about the result of their fishing efforts to shore-based control centers. as well as national and international laws mean that noncompliance can prove very expensive in terms of fines and withdrawal of fishing rights.

In addition, fishermen need accurate position information to locate their assets. Traditional methods often rely on local knowledge and historical patterns of net laying. Transmissions back to the mother ship from free-floating and static devices would improve fishing capabilities.

Figure 8.24   Modern fishing vessels.

This scenario constitutes one of the application directions for the GALILEO system in the L/W layer [79]. In fact, apart from the day-by-day navigation and positioning of vessels, GALILEO can help monitor fish resources. This can also be enhanced with the support of data from the sea and surrounding environments. Certified GALILEO services will allow authorities to confirm that fishing vessels operate only in proper areas.

8.2.2.2   Oil and Gas

A recent and interesting application of GNSS in the water layer is reported in [80]. Due to the extraction of large amounts of oil and water from relatively shallow uncompacted oil reservoirs (since 1932), the area above and adjacent to the Wilmingtom Oil Field in Long Beach, California, began to subside shortly after oil production began. The subsidence threatened to inundate both city and port areas; horizontal surface movements accompanying vertical subsidence caused extensive damage to existing structures, oil wells, pipelines, roadways, and U.S. navy installations. Regular deformation surveys began in the harbor area in 1945 and then extended to the downtown area in 1953. Recently, the city has started a high-precision GPS-based monitoring of the phenomenon, reducing the data collection and processing time and providing timelier subsidence information that can be exploited to predict trends in deformation over the entire area. The program consists of a mobile, biannual GPS campaign as well as the installation of permanent GPS stations.

The oil and gas sector can also benefit from the GALILEO system in many areas [81]. For instance, marine seismic exploration will use the positioning service both for seismic acquisition vessel as well as seismic streamer arrays and gun arrays. This will enhance the safety of drilling activities by enabling high-resolution surveys of the new sites and identification of any geo-morphological or geophysical risk. The positioning of the rig and its anchor-handling vessel will also be improved through GALILEO. Accurate positioning information will be provided during transit and final positioning of tow tugs relative to the rig, the anchoring of semisubmersibles, and any independent drilling rig. The final position of the drilling facility will be determined, as well as the final orientation of the platform to a high degree of accuracy. It is worth mentioning that the trend in the oil and gas sector is to move away from established finds toward remote sites without any local infrastructure. In these areas, satellite positioning and communications are of vital importance. Real-time data transmission combined with position determination enables oil companies to make real-time decisions on drilling operations. Integrity information provide by GALILEO is of paramount importance when approaching the target and preparing to anchor or lower the drilling platform legs.

8.3   Air Layer

In Chapters 4 and 5 large sections were devoted to air applications of satellite navigation systems and the related benefits. Therefore, this chapter is mainly devoted to other applications related to the L/W and S layers. However, it is useful—for the sake of completeness—to summarize the major achievements and perspectives related to the A layer.

A GNSS can be used for air navigation in three basic roles:

- As a *primary* navigation system, hence able to meet all requirements to use certain procedures or to fly in certain airspace without the need for any other navigation system onboard the aircraft. The primary system may include one or more integrated navigation sensors (e.g., GPS with an inertial reference system) [76, 81–86].
- As a *supplemental* system that can be used alone without comparison to another system: however, a primary system that could be used in the event that the supplemental system is not available must be onboard the aircraft.
- As a *multisensor* navigation system that can be used for navigation but only after it has been compared for integrity with a primary system of the aircraft.

8.3.1   Airplanes

When the operational capability of a civil GPS signal arrived in the late 1980s, the civil community within the United States quickly embraced the technology and adopted minimum operational performance standards (MOPS) and TSO for commercial and general aviation (GA) aircraft [35, 87]. It is worth mentioning that the GA term describes all aviation except government and scheduled-airline use. In addition to recreational flying, it includes flight training, shipping, surveying, agricultural applications, air taxis, charter passenger service, corporate flying, emergency transport, firefighting, and more.

Although considered a stand-alone system, most aircraft avionics employing GPS introduced additional receiver software that performs a self-consistency check on the GPS signals. This fundamental layer of integrity (RAIM; see Chapter 4) seemed acceptable as a supplemental form of navigation for all modes of flight from oceanic through nonprecision approach. When the selective availability was removed in 2000, RAIM users quickly received an increased availability of service. However, even without selective availability, civil users could not expect to reach levels of accuracy, integrity, or continuity required for precision-approach operations. Without additional augmentation, CAT. I, II, and III precision-approach capabilities (see Chapter 4) would only be available with ground-based navigation aids [e.g., ILS or microwave landing systems (MLS)].

To reduce the reliance on ground-based navigation aids for CAT. I precision approach in the U.S. NAS, WAAS was created (see Chapter 4). Similarly, other augmentation systems have been developed in Europe (EGNOS), Japan (MSAS), India (GAGAN), and China (SNAS).

Due to the FAA certification criteria and the consequent requirements in terms of aviation safety-related system performance, the WAAS level of service (lateral precision vertical guidance, or LPVC) at IOC falls in between nonprecision approach and CAT. I, meeting integrity requirements in the range between the international standards APV I and II; FOC for WAAS is expected in 2007, whereas WAAS evolution to a CAT. I system will not occur until approximately 2013 [35, 88, 89].

One of the many challenges to systems like WAAS is providing an assured level of integrity to single-frequency civil users. In fact, the ionosphere represents the largest uncompensated error impacting on civil users. To reach CAT. I levels of integrity, SBAS will probably need to reduce significantly the ionosphere-delay-associated uncertainty by adopting dual-frequency code-based measurements. Fortunately, civil users will receive three code-based frequencies as a result of GPS modernization (see Chapter 6). Similarly, the GALILEO SoL service bases its performance on dual-frequency GALILEO signals. As far as FAA and ICAO aviation performance requirements are concerned, GALILEO is envisaged to provide, as a stand-alone system, an APV II mode of service and, by integration with GPS Block II upgrade, up to CAT. I integrity.

The first step to a more useful GNSS for the civil aviation industry is certainly represented by the removal of the selective availability from the GPS signal; in fact, due to the nature of the remaining error sources, no other "fast errors" are inherent in GPS under nominal conditions, hence enabling increased system predictability. Assuming that the DoD does not switch SA on, civil aviation begins to move toward a stand-alone GNSS as a primary means of navigation. This is also allowed by the further important step of the GNSS world: the creation, development, and imminent deployment of a fully civil, highly performing satellite navigation system such as GALILEO.

The interesting equilibrium that these two initiatives have created is the key for civil aviation to aim at fully GNSS-assisted flight operations. The DoD responsibility in keeping SA turned off together with both the U.S. and European capabilities of maintaining the schedule for GPS modernization and GALILEO deployment are all elements of this aviation-friendly integrated vision.

As shown in the U.S. 2001 Federal Radionavigation Systems (FRS) [90], the U.S. DoD "is responsible for developing, testing, evaluating, implementing, operating, and maintaining aids to navigation and user equipment required for national defense." DoD works in conjunction with NIMA, USNO, and the United States Air Force (USAF).

On the other hand, the U.S. DOT is responsible for ensuring safe and efficient civil transportation, acting through the USCG, the FAA, and the SLSDC, in conjunction with other agencies, such as the Federal Highway Administration

(FHWA), ITS-JPO, the Federal Railroad Administration (FRA), the National Highway Traffic Safety Administration (NHTSA), the FTA, the MARAD, the Research and Special Programs Administration (RSPA), the Bureau of Transportation Statistics (BTS), the GPS Interagency Advisory Council (GIAC), and the Civil GPS Service Interface Committee (CGSIC) [90]. In particular, "the FAA has responsibility for development and implementation of radio-navigation systems to meet the needs of all civil and military aviation, except for those needs of military agencies that are peculiar to air warfare and primarily of military concern." The FAA also has responsibility to provide aids to air navigation required by international treaties [90].

In addition, DoD and DOT have roles and tasks in common, stated in a Memorandum of Agreement (MOA) [91], which "recognizes that DoD and DOT have joint responsibility to avoid unnecessary overlap or gaps between military and civil radio-navigation systems and services. Furthermore, it requires that both military and civil needs be met in a manner cost-effective for the U.S. government and civil user community" [90].

Therefore, GPS will most likely remain the key aid of U.S. military navigation and remain under the control of the USAF and DoD. The direct control of WAAS remains, instead, within the FAA and the DOT, although WAAS' main functionality requires GPS SIS (see Chapters 3 and 4). A tighter integration between USAF and FAA activities requires significant interagency cooperation [35].

The density of ground-based reference stations fielded and/or envisaged in WAAS, EGNOS, MSAS, and GAGAN is more than enough to meet the regional requirements of a GNSS integrity channel broadcast from modernized GPS spacecraft. Regional SBAS participation implies that regional authorities provide GNSS operators with access to their network of code and carrier measurements data in real time. In this frame, it is crucial that WAAS, EGNOS, MSAS, and GAGAN provide reliable measurements to the GPS and GALILEO control segments with minimal latency. Furthermore, the operational costs could be shared between GNSS and SBAS providers, although coverage from a single organization could be also viable.

The introduction of GALILEO can also produce positive effects on GPS, rendering available to the GPS control segment additional raw pseudorange measurements from more remote locations. Those measurements, if combined with the regional measurements, can provide a truly global coverage of the GNSS integrity channel to an APV II level. Furthermore, additional ranging sources created from the GALILEO constellation increase the overall robustness to satellite failures in a combined GALILEO-GPS GNSS, yielding global performance at CAT. I levels, assuming the mutual willingness within the GPS and GALILEO systems to provide the necessary exchanges [92].

A future GNSS integrity channel should have the benefit of a reduction in geostationary satellite ownership or leasing costs required to broadcast SBAS signals. In an intermediate period, characterized by the deployment of dual-frequency civil signals, SBAS operations may continue as normal. Once GALILEO and a full constellation of modernized GPS satellites are operational, SBAS operators would have to decide when to interrupt the support for single-frequency user equipment.

Continuity of service for an aircraft approach with vertical guidance, which is defined as the capability to complete an approach once it has begun, can be lost due to either system failure or fault-free conditions. The latter includes radio frequency interference, which interrupts the integrity data channel for extended time intervals. Under such a condition, the aircraft avionics should revert to another integrity channel in view of the aircraft in a timely way. Current SBAS designs guarantee—in FOC—dual coverage of the integrity channel.

The future availability to the A layer of a GNSS integrity channel—with respect to the use of SBAS aid—has additional advantages [35, 92, 93]: Interoperability concerns can be overcome, as the potential use of separate receivers to decode geostationary broadcasts from different augmentation systems in avionics aboard aircraft flying worldwide is avoided; and users at high latitudes often encounter difficulties in receiving signals broadcast from geostationary satellites, whereas a GNSS integrity channel broadcast from MEO with sufficient inclination ensures that integrity data reach users at the most extreme latitudes, including the poles. This would provide pilots, dispatched to the poles on search-and-rescue missions, with the same level of integrity of vertical guidance as that afforded at low and medium latitudes.

Various applications of GPS are reported. For example, an interesting air-related application of GPS is reported in [94]. In 2001 the FAA alerted the San Diego County Airport of the presence of numerous trees intruding into airspace in runway approach fans. Due to the hazard posed to aviation by such obstructions, the FAA mandated that the county Airport Division survey several dozen suspect trees and trim or remove any that exceeded height limitations. The tight timeline (two months) and the absence of preallocated budget for the work were complicated by the county survey section being overbooked and unable to respond in time to meet the deadline. Consequently, the county Airport Division and the Department of Public Works GIS Division devised an inventory plan that used real-time DGPS positioning, GIS mapping, and other measurement instruments to identify trees for trimming or removal. The process required determining the location of each tree, plotting it on a base map, and identifying the elevation of the top of each tree and of the approach fan. GPS was, hence, the key tool for successfully solving the problem.

An additional application proposal for GPS, as a result of the traumatic events of September 11, is to automate the landing of hijacked flights [95]. In this case, a *dead-man-switch* would allow the pilot to turn over navigational control to an onboard GPS-based auto-landing system. The system would broadcast a mayday to air traffic control (ATC), search an onboard database for the nearest suitable airport, alert that airport, receive landing authorization, and land the aircraft there. During these operations, onboard personnel would not be able to regain control of the aircraft. Hence, no amount of violence on board would allow hijackers to use an aircraft as a missile against a target. The FAA proposed two GPS-based systems, WAAS and LAAS, that could enable this antihijacking capability(see Chapter 4).

Additional activities related to GPS exploitation concern the possible integration of GPS with an INS to develop an autonomous navigation system for GA. The study also focuses on a U.S. DOT action plan aimed at maintaining the adequacy of backup systems for each area of operation where GPS is being used for critical transportation applications, due to the identified susceptibility of GPS to unintentional interference, caused by atmospheric effects, signal blockages from buildings, communications equipment, and potential intentional jamming. There are two options to deal with GPS vulnerability: use of adequate terrestrial backup systems (see Chapter 9) during GPS outages caused by interference; and an increase in the robustness of GPS to unintentional and intentional interference sources. This study addresses both options through the evaluation of integrated low-cost GPS/INS, as well as the evaluation accuracy and integrity performance following the loss of GPS for both low-cost and navigation-grade INS. The following potential GA applications of low-cost GPS/INS are identified: attitude and heading reference system (AHRS) for advanced navigation displays; aiding GPS code- and/or carrier-tracking loops with INS to increase the GPS interference margin to mitigate radio frequency (RF) interference; bridging over short-term GPS outages by using inertial navigation guidance; and inertial coasting after the loss of GPS [96].

In addition, a hybrid integrity monitoring solution for precision approach and landing in a GPS environment degraded by RF interference has been designed that includes a spatial environment integrity monitor and a GPS/inertial RAIM solution [97]. An alternative approach to integrity monitoring for aircraft navigation has been also proposed that needs no assumptions on the distribution of the uncompensated pseudorange errors but requires that the measurements set be very redundant. The combined GPS plus GALILEO constellation seems more than adequate for the approach to work [98].

A further A layer application of GPS relates to parachutes. To verify parachuted cargo pallet rigging configuration designs prior to use by operational forces, the U.S. Army needed a practical and rugged instrumentation system giving full x, y, z trajectory from aircraft egress to ground impact. The army also needed to track new troop parachute designs to determine if excessive oscillations occur and how variations of weight and wind affect landing positions [99]. Analysis of these processes required measurements of parachute payload behavior and dynamics during and after exit from the aircraft. This, in turn, required continuous position, velocity, and attitude of pallets. For troopers, velocity, swing, and position were the most important. For both systems, impact location was also needed. The two problems have been approached with two separate instrumentation/data recording packages, both using GPS [99].

A military application of GPS concerns tactical missiles [100, 101]. In particular, HARM (High-speed Anti-Radiation Missile) is an air-to-surface tactical missile. It is a primary weapon system used in military Suppression of Enemy Air Defenses (SEAD), which detects radar signals, identifies their characteristics, and destroys active enemy air defense systems without the precise targeting data required by conventional or precision-guided munitions [102]. HARM does not have a radio link through which it can be controlled or retargeted after launch ("fire and forget" missile type); nonetheless, it can operate in several modes, such as self-protection and seek and destroy. To address these complexities, the International HARM Upgrade Project, sponsored by the German Air Force and Navy, the Italian Air Force, and the U.S. Navy, aimed at developing and installing a precision navigation upgrade (PNU) on the HARM that improves the weapon's effectiveness and highly reduces risk of "fratricide" effects (i.e., hit of installations on its same side in war) and of "collateral damages" (i.e., destruction of nonmilitary targets and killing of civilians). The PNU is equipped with a GPS receiver with an SA/antispoofing module and an inertial measurement unit.

A further military application concerns the delivery of supplies and equipment by aerial transport, which is a critical component of the U.S. military global mobility strategy. In many cases precision airdrops of cargo and supplies are more desirable than landing transport aircraft [103]. It is extremely important to ensure the parachuted cargo lands at the desired site on the ground without being either intercepted by hostile forces or lost. A key factor in calculating the correct dispensing point for the airdrop is accurate information on wind direction and speed versus altitude. It is also highly desirable to be able to track the cargo descent to the ground from the release aircraft or from the ground. This would provide confirmation of a successful drop and allow the receiving ground personnel to locate the item more rapidly, especially in poor visibility conditions. The GPS TIDGET tracking system is ideally suited for providing wind measurements information and tracking/location information on airdropped cargo [104].

Concerning GALILEO, its practical uses are mainly related to commercial air transport, surface movement and guidance control, and leisure. The main aspects related to *commercial air transport* concern free flight, critical flight phases, monitoring, and surveillance. In particular, concerning *free flight*, GALILEO will be used in all the flight phases of commercial aircraft; during en route flight the availability of both GPS and GALILEO will ensure high robustness through the service's redundancy and high reliability. In the future, higher accuracy and service integrity will allow aircraft separation to be reduced in congested airspace to cope with traffic growth. In recent years, scheduled traffic has increased by about 4% per year worldwide. This trend would double the number of flights within 20 years. As a consequence, several intense pressure points and bottlenecks are forming in some areas of the network and, hence, a significant increase in capacity in the short term is needed. Augmentation of the capacity requires increasingly reliable and accurate positioning systems and associated monitoring, provided by adding GALILEO to the existing radio navigation network.

The major need of commercial operators during *critical flight phases*, such as take-off and landing, is to operate in all weather conditions. As a consequence, precision approach is mandatory for a gate-to-gate navigation system. GALILEO, with the aid of ground-based augmentation (local elements; see Chapter 5), will satisfy the needs for precision approach as defined in the aeronautical standards and could replace or complement the navigation infrastructure of airports in regions where the system is inadequate. For example, some airports are not equipped with instrument landing systems. In this case GALILEO offers many benefits for overall safety and optimization of schedules and routes. It will also help increase runway capacity by shortening runway occupancy time. There will result in savings in time and fuel and a reduction in noise.

Concerning *monitoring and surveillance*, it is worth mentioning that position, heading, speed, and time information are needed by air traffic controllers for the continuous management of all aircraft. Some areas of the world lack the appropriate ground infrastructure, including secondary radar and communications links. For example, in the Canary Islands it is available only intermittently, and the radar service is limited and without backup. The standardized transmission from the aircraft of navigation data obtained via GALILEO will lead to advanced systems and techniques for safer air traffic monitoring.

As far as *surface movement and guidance control* is concerned, moving an aircraft on the ground requires as much assistance from air traffic controllers as that required during the flight. The airport may have surface radar, but sometimes the taxi movements are reported manually by the pilots and the aircraft is managed using visual aids only. Several accidents have occurred during this supposedly safe phase. GALILEO—together with its local elements and communications links—will improve the safety of these operations, creating the means for integrated surface movement guidance and control.

Concerning *leisure*, GALILEO and satellite navigation will be available for all kinds of aviation activities, such as ultralight aircraft and recreational flights. The integration of position information and communication links opens up a wide range of applications.

While waiting for the deployment of GALILEO, the relevant features of EGNOS are to be exploited effectively [105].

New EGNOS facilities were inaugurated in Iceland in 2003: this is another important step of the system to support the A layer community, as the monitor stations in Iceland offer EGNOS coverage deep into the north Atlantic region, one of the busiest air traffic routes [105].

In addition, ESA has promoted a study called EGNOS TRAN, for Terrestrial Regional Augmentation Network. The EGNOS TRAN concept is to broadcast parts of the EGNOS data through terrestrial links, hence compensating for the lack of line-of-sight visibility to geostationary satellites [106]. Complementary dissemination of the EGNOS data will be made by means of terrestrial networks such as mobile phone networks, local differential networks, and LORAN-C/Eurofix broadcasts, where the EGNOS satellite signal is not available. Therefore, it is expected that EGNOS services will be provided in areas where signal availability is limited or lacks continuity, such as in difficult terrain, urban areas, and northern latitudes, so as to achieve the same performance characteristics as that provided by using raw EGNOS broadcasts. Thus, EGNOS TRAN could potentially provide improved EGNOS service to an increased number of users and to a larger geographical area than originally envisioned. Among the envisaged applications for EGNOS TRAN are civil aviation, airport surface operations, and ATC.

8.3.2 Other Aircraft

A further example of practical uses of GALILEO concerns *balloons* and *helicopters*. The former can be used in leisure applications of GALILEO, whereas the latter fall within the system's safety-related applications. In particular, the SoL service of GALILEO and the EGNOS signal can be used as a guide to help land search-and-rescue helicopters in bad weather conditions, such as low visibility and fog, where helicopter operations were previously not possible [107]. This will improve the availability of medical helicopter services for severe road accidents, which often occur in bad weather conditions: GALILEO and EGNOS technology will also lead to more reliable use of helicopters in general for those needing urgent hospitalization.

Looking ahead, a further potential application of GNSS in the A layer can be envisaged for high altitude platforms (HAP). These consist of aerial auto-piloted platforms carrying communications relay payloads and operating at altitudes between 15 and 30 km. The latter are considered important building blocks in an integrated network scenario, where terrestrial, satellite, and stratospheric components participate—in a nonconcurrent form—to the optimal provision of advanced communications service for mobile users [108]. The future deployment of HAP-based systems will identify the main areas where GNSS can be most effectively exploited to support this aerial means.

8.4 Space Layer

Low-cost autonomous navigation, onboard maneuver planning, and autonomous constellation control become feasible when a GNSS is employed, leading to a substantial reduction in mission operation costs, mostly related to the labor intensive—overwhelming also in terms of ground personnel requirements—planning and control of a single spacecraft or a constellation from the ground [109]. Installing a GNSS receiver onboard a spacecraft provides the opportunity to use a single lightweight, low-cost sensor for a multitude of functions: *position, velocity, attitude, attitude rate*, and *time*, resulting in reduced cost, power mass, and complexity of the spacecraft. In addition, the replacement of many different sensor devices and their interfaces with a single one would yield a noticeable improvement in system reliability.

The nonspherical shape of the Earth and the nonhomogeneous distribution of its inner mass determine precession and nutation of the satellite orbital plane. These effects can be counterbalanced through periodical corrections on the motions around three satellite axes. As a consequence, the knowledge of the satellite attitude, which is determined by the angles between the satellite body mechanical axes and three reference axes (roll, pitch, and yaw), becomes of paramount importance. All spacecraft—but the simplest ones—adopt some means of active *attitude* control, using actuators such as control momentum gyros, reaction wheels, offset thrusters, and magnetic torque rods. Attitude control is mostly performed by an onboard closed-loop system, in particular in case of LEO spacecraft where daily ground contacts are very restricted in time. Only unusual events, such as momentum unload or spacecraft slew maneuvers, are commanded from the ground, even if the onboard system is also, in these cases, responsible for some automatic control. Closed-loop attitude control necessitates sensor feedback of the vehicle orientation. This task has been traditionally provided either by low-cost sensors, such as magnetometers, horizon sensors, sun sensors, or by more expensive high-performance instruments, including gyroscopes and star trackers.

In recent years, GPS receivers have been exploited effectively in attitude determination of LEO microsatellites [110–117]. The attitude can be achieved through GPS using a multiple (two or more) antenna architecture in an interferometric mode of operation, where the various antennas are located in specific locations onboard the satellite [109].

The technique has proven to be extremely cheap and accurate [112–114]. In particular, the method has been used by loral GLOBALSTAR satellites, which adopted a four-antenna solution with a GPS receiver multiplexing among the antennas to observe the relative carrier phase [109, 118]. The calculation is performed in two steps: the exact phase difference among the carriers at each antenna is determined first; after evaluating the local antenna coordinates, the attitude parameters are evaluated from both the local coordinates and the corresponding satellite frame coordinates [114]. An accuracy better than 1 has been proven using a three-antenna system for microsatellites [114]. System performance mainly depends on the receiver firmware, the GPS satellite geometry, the multipath correction, and the microsatellite dynamics and vibrations. Improvement is expected in the near future by focusing on the real-time resolution of ambiguity and on the real-time dynamic determination of the satellite attitude.

Bandwidth, accuracy, and antenna placement are the key performance related to the usefulness of GNSS-based closed-loop attitude sensor approach [76]. Typical low-precision attitude control systems have bandwidths in the order of a few times the orbit rate and pointing accuracy of 1° to 5°. In the case of high-performance attitude control systems, bandwidths range up to a few Hz, with pointing accuracy of 0.1° or better. The update rate of the GPS attitude determination system is in the 0.1–10-Hz range and the related accuracy strongly depends on antenna locations and data-processing techniques. Both accuracy and bandwidth performance can be improved through a combined use of GPS sensors and gyroscopes.

The operational capability of the GPS attitude sensor in space was demonstrated in June 1993, with the launch of the Air Force-sponsored RADCAL satellite into an 800-km polar orbit [76]. This gravity-gradient stabilized satellite contained two cross-strapped GPS receivers that provided differential carrier measurements to achieve attitude solutions in postprocessing. One of the two receivers failed after six months in orbit.

Use of GPS receivers as an onboard real-time attitude sensor has been considered for other spacecraft, including OAST Flyer, Gemstar, REX-II, Orbcomm, SSTI Lewis, SSTI Clark, and the previously mentioned GLOBALSTAR [76, 119].

An alternative method of attitude determination has been demonstrated for spinning satellites, where a single antenna element is used and the observed cyclic variation in the Doppler offset on the GPS satellite signals is exploited to estimate the current attitude, with an accuracy in the order of 2–3° [109].

As previously outlined, an Earth satellite collecting GNSS (at present, GPS) data with an onboard receiver can compute its state in terms of *position and velocity*. This can be performed in different ways, depending in part on the type of orbit and mission. Tracking and navigation requirements can include: real-time state knowledge and active control during launch and orbit insertion as well as during reentry and landing; real-time relative navigation between vehicles during rendezvous; autonomous station-keeping and near-real-time orbit knowledge for operations and orbit maintenance; rapid postmaneuver orbit recovery; and after-the-fact precise orbit determination for scientific analysis [120–127]. Orbit accuracy requirements range from hundreds of meters (or more) to a few centimeters.

Among the existing tracking systems, only GPS currently meets the most stringent of these needs for the most dynamically unpredictable vehicles. The GPS signal beam-widths extend about 3,000 km beyond the Earth limb, enabling an Earth orbiter below that altitude to receive continuous three-dimensional coverage.

Above 3,000 km, an orbiter begins to lose coverage from GPS. However, because the dynamic model error can be small at high altitude, the dynamic orbit solution can remain strong. By looking downward to catch the signal spillover from satellites on the other side of the Earth, an orbiter can exploit GPS from well above the GPS satellites themselves, out to geosynchronous altitude and beyond. As an alternative, high satellites can carry GPS-like beacons to be tracked from the ground, with GPS satellites serving as reference points (*inverted* GPS technique).

Because the preferred tracking modes and solution techniques differ from low and high orbiters, the application to highly elliptical orbiters (HEO), which may descend to a few hundred kilometers and rise to tens of thousands, is particularly challenging: up- and down-looking GPS combined with ground-based Doppler during the high-altitude phase can provide a particularly strong coverage [76].

Real-time techniques belong to the *direct* GPS orbit determination, where only GPS data collected by the orbiter are used in the solution. As an alternative, for precise after-the-fact solutions, a *differential* GPS approach is proper, where data collected at multiple ground sites are combined with onboard data to reduce major errors.

The potential of GPS to provide accurate and autonomous satellite orbit determination was noted in its early development. Direct GPS-based tracking has been investigated by identifying applications from near Earth to beyond geosynchronous altitudes, GPS tracking of the space shuttle, autonomous near Earth navigation, comparison of GPS, and NASA's tracking and data relay satellite system (TDRSS) for onboard navigation [128–131]. The first reported results from orbital application of direct GPS tracking have been those of the NASA's Landsat-4 experiment, which achieved about 20-m accuracy during the quite short windows of satisfactory GPS visibility [132–134].

Precise orbit determination in the order of several decimeters or better by differential GPS approach were first reported for the TOPEX/Poseidon ocean altimeter mission by exploiting a subdecimeter carrier phase-based technique [135]. Differential techniques have been proposed for both low- and high-altitude satellites for several applications [136–138].

Further examples of space application of the GPS receiver are given next.

A four-antenna, six-channel GPS receiver has been mounted onboard the Gravity Probe B experiment, recently launched after more than 40 years of development [139]. Developed by NASA and Stanford University to test two unverified predictions of Einstein's general relativity theory, Gravity Probe B carries a TANS, for Trimble Advanced Navigation Sensor, vector to generate real-time navigation solutions, where the GPS receiver will help provide orbit trim and raw measurements for ground postprocessing that will produce centimeter-level accuracy *for position, velocity, and attitude* solutions.

The use of GPS is also envisaged onboard the PRIMA (Italian Multi-Applications Reconfigurable Platform), under development at Alenia Spazio in Italy [140]. The orbit control and orbit determination functions, provided by the integrated control system (ICS), are performed, under nominal conditions, through the onboard processing of the GPS sensors data; in degraded conditions (i.e., in case of failure of both GPS units), they are performed propagating degraded performance [140].

The NASA Gravity Recovery and Climate Experiment (GRACE) missions, whose twin satellites began their journey in March 2002, uses high-precision GPS measurements, micron-level intersatellite links (ISL), precision accelerometers, and accurate star cameras to produce gravity field maps of the Earth that are order-of-magnitude more precise than current ones. Furthermore, GRACE also aims to measure temporal variations in the Earth gravity field [141]. It is worth noting that the GPS ground data processing system is one of the key technologies enabling the micron-level ISL. The mission processes GPS data to contribute to the recovery of long-wavelength gravity field, removes errors due to long-term onboard oscillator drift, and aligns measurements between the two spacecraft. The timing functions of GPS for precision orbit determination, in terms of position and velocity as a function of time, enable orbit accuracy better than 2 cm in each coordinate. All GPS data processing for orbit and clock parameters is accomplished by data-driven, automated systems designed for particular spacecraft constellations.

The majority of GPS experiments to date have been performed at orbital altitudes below the GPS satellite constellation; in these orbits, there is good reception from the GPS satellites located above the spacecraft. When operating from orbital location above the GPS spacecraft, GPS signals can only be tracked from satellites that fall within the beam pattern transmitted by the GPS satellites. Early experimenters assumed that GPS satellites could only be viewed from behind the Earth, in the location shown in Figure 8.25. However, a later experiment, performed by the U.S. Air Force on the Falcon Gold satellite, demonstrated that it was possible to track the sidelobes as well as the main lobe of the GPS antenna signals, hence significantly increasing GPS visibility from space users [109, 142].

The experiment has been performed using a NAVSYS TIDGET GPS sensor, collecting snapshots of the GPS information and relaying them to the ground for processing. The TIDGET sensor has been exploited to provide orbit information for a Centaur geostationary orbit transfer vehicle (Figure 8.26). The data collected from the experiment have proven the capability to track GPS satellites through all phases of the transfer orbit and, hence, to receive GPS signals from LEO to GEO satellites [109]. This capability highlights the possible exploitation of GPS to provide orbit information in support of orbit entry operations for LEO, HEO, and GEO satellites.

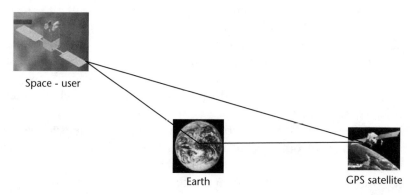

Space - user

Earth   GPS satellite

Figure 8.25   GPS satellite visibility.

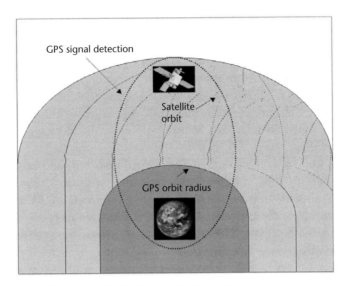

Figure 8.26 GPS signal reception in the Falcon Gold experiment.

A further space-related application of GNSS consists in the exploitation of the GPS time reference for image acquisition in remote sensing applications. It has to be mentioned, however, that the Korea Multi-Purpose Satellite KOMPSAT-1, which undertook a 3-year mission (December 1999–December 2002), experienced cases of anomalies in time synchronization between the onboard time and the GPS time (jitter or short time jumps in the GPS signal received due to, for instance, visibility of the GPS satellite or problems in the GPS receiver), and real-time monitoring of the time error was required to construct remote sensing images for best quality [35]. Therefore, a GPS-based clock synchronization technique employing a refined digital phase lock loop (PLL) circuitry with a front-end processor has been proposed as a fine time-management system for LEO remote sensing satellites, aiming for use in the KOPMSAT-2 mission [35].

The use of GPS in spacecraft is and has been widespread, in particular by the DoD and NASA [19].

Both commercial and military space systems are required to provide accurate, reliable satellite position determination during launch and in orbit. The satellite position during *launch* is currently tracked by ground monitor stations: low-cost tracking systems, based on use of GPS, concur in reducing the overall mission costs, once the GPS user equipment is technologically capable of dealing with the high spinning rate of satellites during launch and orbit entry. Advances in low-cost, digital GPS phased array technology should represent a solution from this point of view.

In 2002 the space shuttle program completed an integration, ground test, and flight test effort to certify a GPS receiver for use on space shuttle orbiters. After a combined use of existing tactical air navigation (TACAN) units, GPS is finally expected to become the sole means for providing accurate navigation data during the shuttle on-orbit and entry phases, also increasing the safety of flight. The space shuttle is not expected to use GPS for attitude determination [143]. A preproduction version of the chosen GPS receiver flew on the Shuttle Endeavor seven times from

December 1993 to May 1996, whereas the first flight of a production receiver was performed in September 1996 [143, 144]. Once the TACAN units are fully replaced, shuttle computers will incorporate GPS state vectors continuously during entry at a rate dependent on the flight phase, although a landing does not need continuous updates. After TACAN replacement, computers will continue to process data from the drag and barometric altimeters. Ground radar tracking will be available as backup to GPS, particularly if any GPS receiver or legacy navigation sensor fails prior to entry.

In addition, GPS attitude determination has been exploited in the International Space Station (ISS) design [145]. The mission-critical requirements for position, velocity, and attitude determination brought to install GPS antennas on the ISS and the crew return vehicle (CRV) in April 2002. Two days after installation, the four-GPS-antenna equipment began using the GPS position and velocity solution as the ISS navigation state. After about 2 weeks, following a check-out period, the ISS GPS capability for attitude determination became operational. The ISS attitude determination filter combines the GPS receiver attitude determination information with data from the ISS rate gyro assembly to produce the ISS attitude solution.

A further space application of GPS concerns formation flying technology. The latter is changing dramatically the way the civil and defense space community conducts missions in space, revolutionizing all future space missions related to earth science, space science, human exploration, defense, and commercial ventures. Collaborative on-orbit experiments and ground-based tools will provide low-cost validation of formation flying hardware and software algorithms. Future missions will rely on space-borne GPS technology and advanced space vehicle autonomy techniques to enable the construction of virtual platforms in space [146].

References

[1]   "National ITS Program Plan," ITS America in Cooperation with the U.S. Department of Transportation, March 1995.

[2]   "White Paper—European Transport Policy for 2010: Time to Decide," Commission of the European Communities, Brussels, Belgium, September 12, 2001.

[3]   "National ITS Program Plan: A Ten-Year Vision," ITS America in Cooperation with U.S. Department of Transportation, January 2002.

[4]   Chowdhury, M. A., and A. Sadek, *Fundamentals of Intelligent Transportation Systems Planning*, Norwood, MA: Artech House, 2003.

[5]   Bin, W., et al., "Study on Adaptive GPS/INS Integrated Navigation System," *Proc. 2003 IEEE Intelligent Transportation Systems*, Vol. 2, Shanghai, China, October 12–15, 2003.

[6]   Prasad, R., and M. Ruggieri, *Technology Trends in Wireless Communications*, Norwood, MA: Artech House, 2003.

[7]   Ferman, M. A., D. E. Blumenfeld, and Dai Xiaowen, "A Simple Analytical Model of a Probe-based Traffic Information System," *Proc. 2003 IEEE Intelligent Transportation Systems*, Vol. 1, Shanghai, China, October 12–15, 2003.

[8]   Liao, Z., "Taxi Dispatching Via GPS," *IEEE Trans. on Engineering Management*, Vol. 48, No. 3, August 2001.

[9]   Commission Communication, "Developing the Trans-European Transport Network: Innovative Funding Solutions—Interoperability of Electronic Toll Collection Systems—Proposal for a Directive of the European Parliament and of the Council on the Widespread

Introduction and Interoperability of Electronic Road Toll Systems in the Community," March 23, 2003.

[10] Prasad, R.., et al. "Vehicular Remote Tolling Services Using EGNOS," *Position Location and Navigation Symposium 2004*, Monterey, CA, April 26–29, 2004.

[11] Ochieng, W. Y., et al., "Integration of GPS and Dead Reckoning for Real Time Vehicle Performance and Emissions Monitoring," *The GPS Solutions Journal*, Vol. 6, No. 4, 2003, pp. 229–241.

[12] Cochran, A., "AHS Communications Overview," *IEEE Conference on Intelligent Transportation System*, Boston, MA, November 9–12, 1997.

[13] Wu, J., et al., "Vehicle to Vehicle Communication Based Convoy Driving and Potential Applications of GPS," *2nd International Workshop on Autonomous Decentralized System*, Beijing, China, 2002.

[14] Lee, J-S., and Pau-Lo Hsu, "An Object-Oriented Design of the Hybrid Controller for Automated Vehicles in an AHS," *IEEE Intelligent Vehicle Symposium*, Vol. 1, Singapore, June 17–21, 2002.

[15] Report: ATIP02.022, *Mobile Internet Navigation Services*, Asian Technology Information Program (ATIP), May 29, 2002.

[16] Styles, J., N. Costa, and B. Jenkins, "In the Driver's Seat," *GPS World*, October 1, 2003.

[17] "Business in Satellite Navigation," ESA/EC, March 2003.

[18] Cramer, M., "On the Use of Direct Georeferencing in Airborne Photogrammetry," *Proc. 3rd Intl. Symp. Mobile Mapping Technology*, Cairo, Egypt, January 3–5, 2001.

[19] Kaplan, E. D., (ed.), *Understanding GPS: Principles and Applications*, Norwood, MA: Artech House, 1996.

[20] El-Rabbany, A., *Introduction to GPS: The Global Positioning System*, Norwood, MA: Artech House, 2002, pp. 141–144, 149–150.

[21] Ash, T., et al., "GPS/Inertial Mapping (GIM) System for Real Time Mapping of Roadways Using WADGPS," *ION Fall Meeting*, Palm Springs, CO, September 1995.

[22] Grejner-Brzezinska, D. A., and C. K. Toth, "Driving the Line—Multi-Sensor Monitoring for Mobile Mapping," *GPS World*, March 1, 2003.

[23] Abwerzger, G., "Georeferencing of Laser Scanner Data Using GPS Attitude and Position Determination," *Proc. 3rd Intl. Symp. Mobile Mapping Technology*, Cairo, Egypt, January 3–5, 2001.

[24] Favey, E., et al., "3-D Laser Mapping and Its Application in Volume Change Detection of Glaciers," *Proc. 3rd Intl. Symp. Mobile Mapping Technology*, Cairo, Egypt, January 3–5, 2001.

[25] Mohamed, A., and R. Price, "Near the Speed of Flight," *GPS World*, March 1, 2002.

[26] Maxfield, H. E., "Recent Developments in Seafloor Mapping Capabilities," *Hydro International*, Vol. 2, No. 1, January/February 1998, pp. 45–47.

[27] Langley, R., et al., "Mapping the Low-Latitude Ionosphere with GPS," *GPS World*, February 1, 2002, based on the paper: "The Low-Latitude Ionosphere: Monitoring Its Behavior with GPS," *14th International Technical Meeting of the Satellite Division of the Institute of Navigation*, Salt Lake City, UT, September 11–14, 2001, pp. 2468–2475.

[28] Kiss, G. K., "The Use of Modern Satellite Systems in the Railway Traffic," *Periodica Polytechnica Ser. Transp. Eng.*, Vol. 28, No. 1–2, 2000, pp. 123–130.

[29] Sauer, S. J., "Burlington Northern Santa Fe GPS Survey Project," *American Railway*, 1997 (quoted by G. K. Kiss, "The Use of Modern Satellite Systems in the Railway Traffic," *Periodica Polytechnica Ser. Transp. Eng.*, Vol. 28, No. 1–2, 2000, pp. 123–130).

[30] Hartberger, M., "Train Influence System by ALCATEL," *Eisenbahnsicherungstechnik*, 1999.

[31] Gruller, R., "NAVSTAR-GPS, the Global Positioning System," *Eisenbahn-Revue*, Vol. 5 and 6, 1997, and Vol. 1 and Vol. 2, 1998, pp. 1–2.

[32] "The Little SANDY is Very Big in Collecting of Data," *Deine Bahn*, Vol. 8, 1999.

[33] Renken, J., "Passenger Security and Satellite-Based Positioning in S-Bahn Berlin," *Siganl+Draht*, Vol. 3, 1996.

[34] "Rail Applications," GALILEO Applications Sheets, May 2003; http://europa.eu.int/comm/dgs/energy_transport/galileo/applications/rail_en.htm.

[35] Rycroft, M., (ed.), *Satellite Navigation Systems: Policy, Commercial, and Technical Interaction*, Boston, MA: Kluwer Academic Publishers, 2003.

[36] Glaus, R., et al., "Precise Rail Track Surveying," *GPS World*, March 1, 2004.

[37] Leahy, F., M. Judd, and M. Shortis, "Measurement of Railway Profiles Using GPS Integrated with Other Sensors," *IEE Vehicle Navigation & Information Systems Conference*, Ottawa, Canada, 1993.

[38] Schiestl, A. J., "A Sense of Place," *GPS World*, March 1, 2004.

[39] Jirawimut, R., et al., "Visual Odometer for Pedestrian Navigation," *IEEE Trans. on Instrumentation and Measurement*, Vol. 52, No. 4, August 2003, pp. 1166–1173.

[40] Jirawimut, R., et al., "A Method for Dead Reckoning Parameter Correction in Pedestrian Navigation System," *IEEE Trans. on Instrumentation and Measurement*, Vol. 52, No. 1, February 2003, pp. 209–215.

[41] "More Autonomy for Blind People Thanks to Satellite Navigation," European Space Agency, June 2003; http://www.esa.int/export/esaSA/SEMVQOS1VED_navigation_0.html.

[42] Ladetto, Q., and B. Merminod, "In Step with INS," *GPS World*, October 1, 2002.

[43] Petrovski, I. G., et al., "Pedestrian ITS in Japan," *GPS World*, March 1, 2003.

[44] "Applications for People with Disabilities," available on GALILEO Web site http://www.europa.eu.int/comm/dgs/energy_transport/galileo/applications/disability_en.htm.

[45] Milnes, K., and T. Ford, "Real-Time GPS FX On-Screen Positioning of Racecars," *GPS World*, September 1, 2001.

[46] Skaloud, J., et al., "With Racing Heart," *GPS World*, October 1, 2001.

[47] http://www.suunto.com.

[48] Lambert, M., and R. Santerre, "Performance Monitoring with RTK GPS," *GPS World*, February 1, 2004.

[49] Lohnert, E., et al., "Wireless in the Alps," *GPS World*, March 1, 2004.

[50] http://www.europa.eu.int/comm/dgs/energy_transport/galileo/applications/leisure_en.htm.

[51] Mikkola, C., "GPS Wherever You Go," *GPS World*, August 1, 2003.

[52] Kee, C., et al., "Centimeter-Accuracy Indoor Navigation Using GPS-Like Pseudolites," *GPS World*, November 1, 2001.

[53] Kee, C., et al., "Development of Indoor Navigation System Using Asynchronous Pseudolites," *Proc. of ION GPS-2000*, Salt Lake City, UT, September 19–22, 2000.

[54] Isshiki, H., et al., "Theory of Indoor GPS by Using Reradiated GPS Signal," *ION NTM 2002*, San Diego, CA, January 2002.

[55] van Diggelen, F., "Indoor GPS Theory & Implementation," *IEEE Position, Location & Navigation Symposium*, Palm Springs, CA, 2002.

[56] van Diggelen, F., and C. Abraham, "Indoor GPS: The No-Chip Challenge," *GPS World*, September 1, 2001.

[57] Vittorini, L. D., and B. Robinson, "Optimizing Indoor GPS Performance," *GPS World*, November 1, 2003.

[58] Lenain, R., et al., "A New Nonlinear Control for Vehicle in Siding Conditions: Application to Automatic Guidance of Farm Vehicles Using RTK GPS," *Proc. of the 2004 IEEE International Conference on Robotics & Automotion*, New Orleans, LA, April 2004.

[59] "Agriculture and Fisheries," GALILEO Application Sheets, June 2002; http://www.europa.eu.int/comm/dgs/energy_transport/galileo/index_en.htm.

[60] Deimert, K., and R. Mailler, "A Good Host," *GPS World*, January 1, 2004.

[61]   "GALILEO—The European Programme for Global Navigation Services," ESA Publication Division, Noordwijk, the Netherlands, May 2002.

[62]   Stergiou, P., and D. Kalokitis, "Keeping the Lights On," *GPS World*, November 1, 2003.

[63]   Kijewski-Correa, T., and A. Kareem, "The Height of Precision," *GPS World*, September 1, 2003.

[64]   Wong, K.-Y., K.-L. Man, and W.-Y. Chan, "Real Time Kinematic Spans the Gap," *GPS World*, July 1, 2001.

[65]   Reigber, C., et al., "Water Vapor Monitoring for Weather Forecasts," *GPS World*, January 1, 2002.

[66]   Langley, R., D. Kim, and S. Kim, "Shipyard Giants," *GPS World*, September 3, 2002

[67]   Menard, R. J., and J. L. Knieff, "GPS at Ground Zero," *GPS World*, September 2, 2002.

[68]   Bahr, D., F. Schöttler, and C. Schlums, "Save Your Breath," *GPS World*, May 2, 2002.

[69]   Pomogaev, O., "Egypt's Hidden Depths," *GPS World*, November 1, 2002.

[70]   Scott, L., and D. E. Denning, "Geo-Encryption," *GPS World*, April 1, 2003.

[71]   Enge, P. K., M. Ruane, and L. Sheynblatt, "Marine Radiobeacons for the Broadcast of Differential GPS Data," *Proc. IEEE PLANS*, 1986.

[72]   Sennott, J. W., and I. S. Ahn, "Simulation of Optimal Marine Waypoint Steering with GPS, LORAN-C, and RACON Sensor Options," *Proc. ION NTM*, Santa Barbara, CA, January 1988.

[73]   Amerongen, J., et al., "Model Reference Adaptive Autopilots for Ships," *Automatica*, Vol. 11, 1975.

[74]   Astrom, K. J., and C. G. Kallstrom, "Identification of Ship Steering Dynamics," *Automatica*, Vol. 12, 1976, p. 9.

[75]   Fung, P., and M. J. Grimble, "Dynamic Ship Positioning Using a Self-Tuning Kalman Filter," *IEEE Trans. on Automatic Control*, Vol. 28 Issue 3, March 1983, pp. 339–349.

[76]   Parkinson, B. W., and J. J. Spilker Jr. (eds.), "Global Positioning System: Theory and Applications," *Progress in Astronautics and Aeronautics*, American Institute of Aeronautics and Astronautics, Vols. 163 and 164, 1996.

[77]   Hartberger, A., "Introduction to the US Coast Guard Differential GPS Program," *Proc. IEEE PLANS*, Monterey, CA, March 2002.

[78]   Bachmann, E. R., et al., "Evaluation of An Integrated GPS/INS System for Shallow-Water AUV Navigation (SANS)," *Proc. IEEE PLANS*, Atlanta, GA, 1996.

[79]   "Agriculture and Fisheries," GALILEO Applications, ESA/EC, June 2002.

[80]   Rutledge, P. D., et al., "GPS Monitors Oilfield Subsidence," *GPS World*, June 4, 2004.

[81]   Hogle, L., "Investigation of Potential Applications of GPS for Precision Approaches," *Navigation*, Vol. 35, No. 3, 1988.

[82]   Swider, R., R. Loh, and C. Shively, "Overview of the FAA's Differential GPS CAT. III Program," *Proc. of the Symposium on Worldwide Communications, Navigation, and Surveillance*, Reston, VA, April 1993.

[83]   Till, R. D., V. Wullschleger, and R. Braff, "GPS for Precision Approaches: Flight Testing Results," *Proc. Institute of Navigation Annual Meeting*, Cambridge, MA, June 1993.

[84]   Pilley, H. R, and L. V. Pilley, "Collision Prediction and Avoidance Using Enhanced GPS," *Proc. ION GPS-92*, Albuquerque, NM, September 1992.

[85]   Nilsson, J., "Time-Augmented GPS Aviation and Airport Applications in Sweden," *GPS World*, April 1992.

[86]   Massoglia, P. L., M. T. Pozesky, and G. T. Germana, "The Use of Satellite Technology for Oceanic Air Traffic Control," *Proc. of the IEEE*, Vol. 77, No. 11, 1989.

[87]   Donohue, G., "Vision on Aviation Surveillance Systems," *Proc. IEEE International Radar Conference*, Atlanta, GA, 1995.

[88]   Davis, M.A., "WAAS is Commissioned," *SatNav News*, Vol. 21, November 2003, pp. 1–2.

[89] McArthur, P., "Clear Skies Ahead," *IEEE Spectrum*, Vol. 39, No. 1, January 2002, pp. 79–81.

[90] "2001 Federal Radionavigation Systems," U.S. Department of Defense, U.S. Department of Transportation, 2001.

[91] "Memorandum of Agreement Between the Department of Defense and the Department of Transportation on Coordination of Federal Radionavigation and Positioning Systems Planning," January 19, 1999.

[92] Fyfe, P., et al., "GPS and GALILEO—Interoperability for Civil Aviation Applications," *Proc. ION-GPS 2002*, Portland, OR, September 2002.

[93] Bretz, E. A., and J. Kumagal, "Aerospace & Military," *IEEE Spectrum*, Vol. 37, No. 1, January 2000, pp. 98–102.

[94] Mc Camic, F., "Cleared for Take-Off," *GPS World*, July 1, 2003.

[95] Luccio, M., "GPS and Aviation Safety," *GPS World*, October 1, 2002.

[96] Soloviev, A., and F. van Graas, "Application for General Aviation," *GPS World*, March 1, 2004.

[97] Gols, K. L., and A. K. Brown, "A Hybrid Integrity Solution for Precision Landing and Guidance," *Proc. IEEE PLANS*, Monterey, CA, April 2004.

[98] Misra, P., and S. Bednarz, "Navigation for Precision Approaches," *GPS World*, April 1, 2004.

[99] Strus, J., et al., "15 Tons 1500 Feet 4 Gs—Airdrop Behavior of Parachuted Cargo Pallets," *GPS World*, April 1, 2003.

[100] Mc Neff, J., "GPS in Aerial Warfare—A Matter of Precision," *GPS World*, September 1, 2003.

[101] Russel, M., and J. M. Hasik, "The Precision Revolution: GPS and the Future of Aerial Warfare," Naval Institute Press, June 2002.

[102] Loffler, T., and J. Nielson, "More Precise HARM—GPS/INS Integration to Improve Missile," *GPS World*, May 1, 2002.

[103] Caffery, D. E., and A. Matini, "Test Results of GPS Dropwindowsonde and Application of GPS in Precision Airdrop Capability Using the TIDGET GPS Sensor," *Proc. ION-GPS-96 Conference*, Kansas City, MO, September 1996.

[104] Vella, M. R., "Precision Navigation in European Skies," *IEEE Spectrum*, Vol. 40, No. 9, September 2003, p. 16.

[105] "Iceland Part of Europe's First Satellite Navigation System," *ESA News*, December 2003, http://www.esa.int.

[106] Redeborn, J., et al., "Applicazioni Aeronautiche e GIS di EGNOS TRAN/Aeronautical and GIS Applications of EGNOS TRAN," *Atti Istituto Italiano della Navigazione*, No. 173, December 2003, pp. 41–63.

[107] "GALILEO Applications—Aviation," ESA/EC, No. 007, October 2002.

[108] Cianca, E., and M. Ruggieri, "SHINES: A Research Program for the Efficient Integration of Satellites and HAPs in Future Mobile/Multimedia System," (invited paper), *Proc. WPMC*, Yokosuka, Japan, October 2003, pp. 478–482.

[109] Brown, A., "Space Applications of the Global Service Positioning and Timing Service," *Proc. Richard H. Battin Astrodynamics Conference*, College Station, TX, March 2000, Paper No. AAS-00-269.

[110] Joihi, C. A., O. Eric, and P. H. Jonathan, "Experiments in GPS Attitude Determination for Spinning Spacecraft with Non-Aligned Antenna Arrays," *Proc. 9th International Technical Meeting of the Satellite Division of the Institute of Navigation*, Nashville, TN, September 1998, pp. 1743–1750.

[111] Purivigraipong, S., M. J. Unwin, and Y. Hashida, "Demonstrating GPS Attitude Determination from UpSat-12 Flight Data," *Proc. 11th International Technical Meeting of the Satellite Division of the Institute of Navigation*, Salt Lake City, UT, September 2000.

[112] Susan, F. G., "Attitude Determination and Attitude Dilution of Precision (ADOP) Results for International Space Station Global Positioning System (GPS) Receiver," *Proc. 11th International Technical Meeting of the Satellite Division of the Institute of Navigation*, Salt Lake City, UT, September 2000, pp. 1995–2002.

[113] Unwin M. J., et al., "Preliminary Orbital Results From the SGR Space GPS Receiver," *Proc. 10th International Technical Meeting of the Satellite Division of the Institute of Navigation*, Nashville, TN, September 1999, pp. 849–855.

[114] Dai, L., K. Voon Ling, and N. Nagrayan, "Attitude Determination for Microsatellite Using Three-Antenna Technology," *Proc. IEEE Aerospace Conference*, Big Sky, MT, March 2004.

[115] Lu, G., et al., "Attitude Determination in a Survey Launch Using Multi-Antenna GPS Technologies," *Proc. National Technical Meeting, The Institute of Navigation*, Alexandria, VA, 1993, pp. 251–260.

[116] Lu, G., et al., "Shipborne Attitude Determination Using Multi-Antenna GPS Technologies," *IEEE Trans. on Aerospace and Electronic Systems*, 1994, pp. 1053–1058.

[117] Hoyle, V. A., et al., "Low-Cost GPS Receivers and Their Feasibility for Attitude Determination," *Proc. National Technical Meeting, the Institute of Navigation*, San Diego, CA, 2002, pp. 226–234.

[118] Sacchetti, A., "GPS for Orbit and Attitude Determination: Hardware Design and Qualification Plan for Spaceborne Receiver," *Proc. ION-GPS-94*, Salt Lake City, UT, September 1994.

[119] Cohen, C. E., et al., "Space Flight Tests of Attitude Determination Using GPS," *International Journal of Satellite Communications*, Vol. 12, September–October 1994, pp. 427–433.

[120] Axelrad, P., and B. W. Parkinson, "Closed Loop Navigation and Guidance for Gravity Probe B Orbit Insertion," *Navigation*, Vol. 36, 1989, pp. 45–61.

[121] Hesper, E. T., et al., "Application of GPS for Hermes Rendezvous Navigation," Spacecraft Guidance, Navigation and Control Systems, ESA, 1992, pp. 359–368.

[122] Axelrad, P., and J. Kelley, "Near-Earth Orbit Determination and Rendezvous Navigation Using GPS," *Proc. IEEE PLANS '86*, Las Vegas, NV, November 1986, pp.184–191.

[123] Chao, C. C., et al., "Autonomous Station-Keeping of Geo-Synchronous Satellites Using a GPS Receiver," *Proc. AIAA Astrodynamic Conference*, Hilton Head, NC, August 1992, AIAA CP-92-4655, pp. 521–529.

[124] Lichten, S. M., et al., "A Demonstration of TDRS Orbit Determination Using Differential Tracking Observables from GPS Ground Receivers," *Proc. 3rd AIAA Spaceflight Mechanics Meeting*, Pasadena, CA, February 1993, Paper No. AAS 93–160.

[125] Yunck, T. P., et al., "Precise Tracking of Remote Sensing Satellites with the Global Positioning System," *IEEE Trans. on Geoscience and Remote Sensing*, Vol. 28, 1990, pp.108–116.

[126] Schreiner, W.S., et al., "Error Analysis of Post-Processed Orbit Determination for the Geosat Follow-On Altimetric Satellite Using GPS Tracking," *Proc. AIAA Astrodynamic Conference*, Hilton Head, NC, August 1992, Paper no. AIAA CP-92-4435, pp. 124–130.

[127] Munjal, P., W. Feess, and M. P. V. Ananda, "A Review of Spaceborne Applications of GPS," *Proc. of ION GPS '92*, Washington, D.C., September 1992, pp. 813–823.

[128] Farr, J.E., "Space Navigation Using the Navstar Global Positioning System," *Proc. Rocky Mountain Guidance and Control Conference*, Keystone, February 1979, Paper no. AAS 79–001.

[129] Van Leeuwen, A., E. Rosen, and L. Carrier, "The Global Positioning System and Its Applications in Spacecraft Navigation," *Navigation*, Vol. 26, 1979, pp. 204–221.

[130] Kurshals, P. S., and A. J. Fuchs, "Onboard Navigation: The Near-Earth Options," *Proc. Rocky Mountain Guidance and Control Conference*, Keystone, CO, February 1981, pp. 67–89.

[131] Jorgensen, P., "Autonomous Navigation of Geosynchronous Satellites Using the Navstar Global Positioning System," *Proc. National Telesystems Conference, NTC '82*, Galveston, TX, November 1982, pp. D2.3.1–D2.3.6.

[132] Wooden, W. H., and J. Teles, "The Landsat-D Global Positioning System Experiment," *Proc. AIAA Society Conference*, Danvers, MA, August 1980, AIAA CP-80-1678.

[133] Heuberger, J., and L. Church, "Landsat-4 Global Positioning System Navigation Results," *Proc. AIAA Astrodynamic Conference*, Part I, Lake Placid, NY, August 1983, AAS Paper 83-363, pp. 589–602.

[134] Fang, B. T., and E. Seifert, "An Evaluation of Global Positioning System Data for Landsat-4 Orbit Determination," *Proc. AIAA Aerospace Science Meeting*, Reno, NV, January 1985, AIAA CP-85-0286.

[135] Ondrasik, V. J., and S. C. Wu, "A Simple and Economical Tracking System With Sub-Decimeter Earth Satellite and Ground Receiver Position Determination Capabilities," *Proc. 3rd International Symposium on the Use of Artificial Satellites for Geodesy and Geodynamics*, Ermioni, Greece, September 1982.

[136] Ananda, M. P., and M. R. Chernick, "High-Accuracy Orbit Determination of Near Earth Satellites Using Global Positioning System (GPS)," *Proc. IEEE Plans '82*, Atlantic City, NJ, December 1982, pp. 92–98.

[137] Wu, S. C., "Orbit Determination of High-Altitude Earth Satellites: Differential GPS Approaches," *Proc. 1st International Symposium on Precise Positioning With the Global Positioning System*, Rockville, MD, April 1985.

[138] Yunck, T. P., W. G. Melbourne, and C. L. Thornton, "GPS-Based Satellite Tracking System for Precise Positioning," *IEEE Trans. on Geoscience and Remote Sensing*, Vol. 23, July 1985, pp. 450–457.

[139] "GPS Onboard the Gravity Probe B," *GPS World*, April 26, 2004.

[140] Ritorto, A., et al., "PRIMA Capabilities for DAVID Communication Experiment," *Proc. IEEE Aerospace Conference*, Big Sky, MT, March 2002, Paper No. 4.0902.

[141] Dunn, C., et al., "Instrument of Grace—GPS Augments Gravity Measurements," *GPS World*, February 1, 2003.

[142] Powell, T., et al., "GPS Signals in a Geosynchronous Transfer Orbit: Falcon Gold Data Processing," *Proc. the Institute of Navigation National Technical Meeting*, San Diego, CA, January 1999, pp. 575–585.

[143] Goodman, J., "Parallel Processing—GPS Augments TACAN in the Space Shuttle," *GPS World*, October 1, 2002.

[144] Goodman, J., "Space Shuttle Navigation in the GPS Era," *Proc. of the Institute of Navigation National Technical Meeting*, Long Beach, CA, January 2001, pp. 709–724.

[145] Gomez, S. F., "Flying High—GPS on the International Space Station and Crew Return Vehicle," *GPS World*, June 1, 2002.

[146] Leither, J., et al., "Formation Flight in Space—Distributed Spacecraft Systems Develop New GPS Capabilities," *GPS World*, February 1, 2002.

CHAPTER 9

Integration with Existing and Future Systems

9.1 Introduction

As shown in Chapter 8, satellite-based navigation systems have found application in many fields and even more are expected in the future. Nonetheless, in some situations the user receiver may be unable to track the required minimum number of satellites. This, for instance, may be the case in urban canyons or open-pit mines.

The signal-obstruction problem, however, has been successfully overcome for GPS by integration with other positioning systems or sensors. Augmentation of GNSS is not limited to sensor integration. In fact, systems can be augmented with computer-based tools, such as GIS (see Chapter 8), for efficient data collection analysis. In this context, the word "integration" is intended as an aid to GNSS to improve its performance and remove some of its intrinsic limitations (the *passive integration* approach). Even GPS augmentations in the form of barometric altimeter aiding and clock coasting have been reported as a means of improving availability of the navigation RAIM (see Chapter 4) functions.

Integration, however, can also be intended in the opposite way, as an intrinsic aid to other systems to help them improve their performance and effectiveness (the *active integration* approach). One example is cellular communications systems, which are somehow limited in their ability to determine precisely the location of the origin of a call [1, 2]. This limitation can be critical in emergency situations. For instance, about one-half of all 911 emergency calls in the United States come from cellular phones and in many of these cases the precise location is not available. This makes it very difficult for an operator to dispatch effective assistance. The continuing positive progress in mobile networks is driving a high level of integration with localization (positioning) systems to provide more advanced, dependable, and effective services to the end user.

This chapter moves from these possible meanings of *integration* to provide a broad picture of various system typologies and technologies that can be integrated with GNSS. In particular, two system types are identified: *other navigation* systems and *communications* systems. Among the navigation systems, three are considered: terrestrial, satellite, and mechanical systems. In the case of communications systems, both wireless and all-satellite networks are analyzed together with emergency-dedicated networks.

9.2 Integration with Other Navigation Systems

9.2.1 Radio-Navigation Systems

9.2.1.1 Terrestrial Systems

The terrestrial radio-navigation systems can be classified depending on their maximum range of operation. This distance strictly depends on the frequency range used for radio-wave transmission. Table 9.1 shows this dependency, highlighting how the maximum distance decreases with increasing transmission frequency because of the different radio-wave propagations: ground-wave propagation and sky-wave for low frequencies and direct-wave propagation (LOS link) for high frequencies [3, 4].

In this context, terrestrial radio-navigation systems can be grouped into three main classes:

- *Long-range systems* (LRS): specifically for oceanic navigation (en route oceanic phase of flight and oceanic maritime navigation) and characterized by low accuracy range (500m–10 km);
- *Short-/middle-range systems* (SRS/MRS): specifically for en route domestic phase of flight and coastal maritime navigation and characterized by a maximum accuracy range of 100–500m;
- *Landing systems* (LS): specifically for approach phase of flight and characterized by high or very high accuracy, respectively, with an accuracy range of 100–10m and 10–1m or higher.

Table 9.2 shows the main ground-based radio-navigation systems currently in operation per each class.

Table 9.1 Dependency of Maximum Operability Range on Transmission Frequency

Frequency	Propagation	Distance Range
Very low frequency (VLF)	Ground-wave	Up to 2,000 km
Low frequency (LF)	and	
	Sky-wave	
Very high frequency (VHF)	Direct-wave	50–400 km
Ultra high frequency (UHF)		

Table 9.2 Terrestrial Radio-Navigation Systems

LRS	SRS/MRS	LS	Transmission Frequency Range
LORAN-C			LF
	ADF/NDB		LF, MF
	VOR		VHF
		ILS	VHF, UHF
	DME		UHF
	TACAN		UHF

All the navigation systems shown in Table 9.2 have the potential for integration with GNSS, although true integration at the system level has so far been conceived only for LORAN-C [5–8]. In particular, in the medium-term, when GNSS will represent the primary aerial, maritime, and road navigation systems, all these terrestrial navigation aids will represent valid backups [6].

The next section is devoted to the main features of LORAN-C, highlighting the integration concept in terms of both the passive and active approach.

Concerning other terrestrial radio-navigation systems, it is beyond the purpose of this chapter to describe systems other than the ones whose integration with GNSS is either already developed or underway. Nonetheless, to stimulate further thoughts about the integration with GNSS, what follows is a short—and necessarily limited—description of automatic direction finder/nondirectional beacon (ADF/NDB), DME, TACAN, and ILS. All these systems are mostly employed in the airborne environment; they require a constant visibility of related radio-beacon stations and, hence, a terrestrial worldwide radio-beacon network is strictly required to ensure a global navigation service. This is the main weakness of this approach, which brings a significant decrease in the efficiency and safety of the air traffic environment because air users have to follow predefined routes according to the radio-beacon locations. For these reasons, as outlined in Chapter 10, current air traffic management (ATM) studies are focusing on the future concept of Free Flight, which aims to set the air user free from ground facilities and where GNSS-based technologies and their wide coverage capabilities will play a key role.

The ADF/NDB consists of two subsystems: the automatic direction finder receiver and the nondirectional beacon transmitter station. The ADF receiver, by processing the radio signal broadcast from an NDB station, allows the user to determine the direction of the source of the transmitted signal. ADF receivers use two antennae—one with the axis parallel to the longitudinal axis of the user (i.e., the aircraft in aeronautical applications), and the other with the axis transverse to the first one—to intercept the signal and then determine the direction of that signal. The directional information, displayed on the user onboard instrumentation, is expressed in terms of *azimuth* angle with a maximum error of 5. The azimuth γ defines the clockwise angle between the north line (direction) (with respect to the user) and the line connecting the NDB station and user, as shown in Figure 9.1.

The NDB station consists of a radio transmitter, an antenna coupling device, and an antenna. The NDBs are located in fixed points suitable for en route and approach phases of flight. On demand or with a defined time interval, the NDB broadcasts in every direction the signal to which the ADF points. The signal is transmitted in the 190- to 1,800-kHz frequency range (LF and MF radio bands). Moreover, each NDB is characterized by an identification code of three letters transmitted to users in Morse code [3, 4].

The azimuth can also be measured by using the VOR system, which provides directional information with a maximum error of 3, which is better than 5f rom the NDB. The VOR transmission frequency ranges from 108.0 to 117.95 MHz. It is based on the transmission by the VOR facility of two signals at the same time. One signal is constant in all directions, whereas the other is rotated about the station with a velocity of 30 rounds per minute (rpm). The two signals are transmitted to achieve a phase shift equal to the azimuth angle with respect to

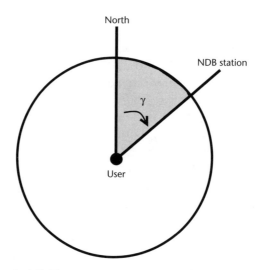

Figure 9.1 Azimuth angle definition.

user mean (see Figure 9.1). The airborne equipment receives both signals, evaluates their phase shift by differencing the two signals, and, hence, determines the azimuth information [3, 4].

The DME system is largely used in the airborne environment to perform range measurement, more specifically, to determine the distance between the mobile user and the DME ground station. The latter is usually colocated with a VOR ground facility (DME/VOR station). The DME operates in the UHF range (from 960 to 1,215 MHz) and is based on time measurement. The DME user transponder determines its distance from a land-based DME transponder by sending two pulses with a random time interval between them. The ground transponder echoes the pulses with a fixed delay t_d of 50 μs. The DME user receiver then intercepts these echoes and measures the elapsed time t_e between the pulse transmission and the pulse reception [3, 4]. The distance d is then obtained as shown in (9.1):

$$d = \left(t_e - t_d\right)c\,\frac{1}{2} \tag{9.1}$$

where c is the speed of light in vacuum.

The *tactical air navigation* (TACAN) system is the military counterpart of VOR and DME. TACAN provides both measures of azimuth and distance, respectively, with a method similar to the basic principles of the VOR and DME. The TACAN ground facility is often co-located with VOR ones (VORTAC stations) and is fully compatible with the DME so that, in the presence of a VORTAC station, the DME is substituted by TACAN. In this case, in fact, civil users determine azimuth and range information by using, respectively, VOR and TACAN facilities of the VORTAC station [3, 4].

ILS is the most common and widespread precision approach system [3, 4]. ILS provides an approach path for exact alignment and descent of an aircraft on final approach to runway. The ILS consists of three functional parts:

• *Guidance information*: localizer, glide slope;

- *Range information*: marker beacon, DME;
- *Visual information*: approach lights, touchdown and centerline lights, runaway lights.

The localizer equipment is a VHF transmitter and is the primary component of the ILS. The localizer has 40 channels available within the frequency range of 108.10 to 111.95 MHz, with 50-kHz mutual spacing. The transmitter and antenna are located about 300m beyond the runway end. The localizer provides lateral guidance, allowing a pilot to identify the centerline of the runway. The transmitter identifies a vertical plane (*localizer* plane) above the runway by modulating two signals with vertical fan-shaped patterns, respectively, on the left and right of the extended centerline of the runway. The left signal is modulated at 90 Hz and identifies the so-called "yellow" area, whereas the right signal is modulated at 90 Hz and identifies the so-called "blue" area. The signal patterns, and, hence, the two areas, overlap in the center and this overlap provides the on-track signal, allowing pilots to line up with the runway center.

The vertical guidance during the approach phase of flight is provided by glideslope equipment, which allows pilots to identify the ideal approach track. The glideslope equipment consists of a UHF transmitter and an antenna system, located 750 to 1,250 ft down the runaway from the threshold and offset 400 to 600 ft from the runaway centerline. The glideslope, like the localizer, has 40 channels, since it operates within the frequency range of 329.15 to 335.00 MHz with 150-kHz spacing between each frequency. The basic principle of operation of the glideslope is similar to that of the localizer. The transmitter, in fact, identifies an equi-signal zone over the optimum glideslope path by generating two overlapping beams modulated at 90 Hz and 150 Hz, respectively, above and below this path. The overlap area identifies the glideslope plane that has a thickness of 0.7 above and below the optimum glideslope path. The inclination of the glideslope depends on the specific airport environment, although the typical inclination is 3° above the horizontal plane.

Measurements of distance from the runaway are provided by the marker beacons. These beacons provide distance information by identifying predetermined points along the approach track. ILS identifies three markers: outer marker, middle marker, and back marker. The markers are low-power transmitters that transmit vertically with an operation frequency of 75 MHz. In some cases, markers are replaced by DME units.

LORAN-C

The LOng RAnge Navigation system started working in the 1970s with an official declaration by the United States and Canadian governments. The system was developed in about 30 years (between 1945 and the 1970s) by the United States in cooperation with the Soviet Union (the equivalent Soviet system is called Chayka) primarily for military use for both marine and air navigation. Currently, LORAN-C provides complete coverage of North America (coastal waters and continental areas of the United States, most of Alaska, and Canadian waters), European coastal areas, and much of East Asia, including, more specifically, the North Atlantic, North Pacific and Mediterranean, and Bering Seas.

The ground segment includes several worldwide stations (LORSTAs) that periodically transmit a group of eight pulses with a center frequency of 100 kHz (LF band). LORSTAs are grouped into *chains*. Each chain is characterized by a unique group repetition interval (GRI), varying from 50 to 100 ms, and contains one master station and two to five secondary ones (the typical configuration includes three secondary stations). Within a given chain, each transmitter emits its group of pulses every GRI with proper synchronization between master and secondary stations. The LORAN-C positioning concept is based on the *hyperbolic method*. The LORAN user receiver, in fact, tracks these signals and determines the difference in the time-of-arrival between the pulse groups from the master station and each secondary station. For each couple of stations, master-secondary, the measured time difference defines a hyperbolic line of position (LOP) for the user. The intersection of two such LOPs defines the user position (see Figure 9.2) [3, 4, 9].

LORAN-C has an absolute accuracy (i.e., the accuracy with which a user can estimate his position with respect to the Earth coordinates) of approximately 400m. The accuracy is limited by the ground-wave propagation effects (the sky-wave propagation is not suitable for positioning because of the uncertainties of the ionospheric effects), which essentially results in random and bias errors.

As previously outlined, LORAN-C performance can be improved by integrating with GNSS. In this context, some studies have investigated the possibility of employing GPS capabilities in order to:

- Perform cross-chain synchronization;
- Calibrate LORAN propagation errors;
- Combine pseudorange measurements.

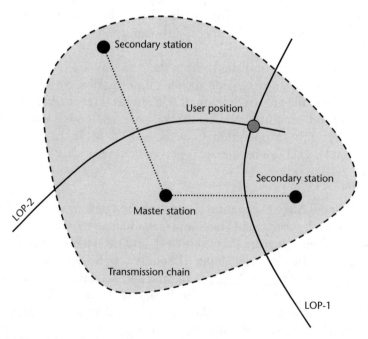

Figure 9.2   LORAN-C hyperbolic method.

The first study examined the time transfer capability provided by GPS to synchronize stations of different transmitting chains [5]. LORAN-C, in fact, provides a precise synchronization only for signals of the same chain, allowing receivers to compute time differences only between pulses of a given chain. This limits system coverage significantly, particularly when the receiver is unable to track completely the signals of a chain. This, for instance, happens when a transmitter fails or in case of particularly noisy environments. This limitation can be overcome by using LORAN-C/GPS integrated receivers. In this case, synchronization through the GPS signal enables the computation of "cross-chain" time differences. This technique allows the LORAN-C coverage to be increased.

The second active integration approach discussed here aims at calibrating LORAN propagation uncertainties to improve LORAN absolute accuracy [5]. As previously highlighted, the positioning concept of LORAN-C is based on time arrival measurements, and the main system error source is due to signal propagation effects. The total travel time (t_{TOT}) of a LORAN signal can be modeled as follows:

$$t_{TOT} = t_{PF} + t_{SF} + t_{ASF} \qquad (9.2)$$

where t_{PF} is the travel time of the LORAN signal moving at the speed of light in air with no boundary effects, and t_{SF} and t_{ASF} are the additional times needed to cross all-water and all-land paths, respectively. Determining t_{ASF} is the most critical, since its figure strictly depends on the structure of the traversed terrain (in terms of conductivity coefficient), the distance traveled by LORAN pulse, and weather conditions. An accurate estimation of t_{ASF} can be obtained by using GPS position fixes, hence allowing users to form fixed correction tables that can be used through proper software by LORAN receivers to estimate precisely t_{TOT} of the signals. This technique is called *GPS calibration* and allows the absolute accuracy of LORAN-C to be improved by calibrating the propagation uncertainties, as GPS/LORAN observations in the Gulf of Maine have tested.

The third study considered an active integration approach at the information level, where GPS capabilities were used to improve significantly LORAN availability, reliability, and absolute accuracy [5]. More specifically, considering the hybrid user equipment, this technique enables one to enhance LORAN performance by combining LORAN measurements with GPS pseudoranges. The data can be combined in two ways: GPS pseudoranges and LORAN time differences, or GPS pseudoranges and LORAN pseudoranges. To integrate properly GPS and LORAN data, and thus achieve the maximum benefits, it is necessary to avoid the unknown time offset between the two navigation systems. Therefore, a transmission time synchronization of all LORAN transmitters is needed, and this requirement is not currently met by LORAN-C. In addition, the success of this technique is bound to the GPS signal availability and, hence, in areas where GPS coverage is weak the improvements are not so evident.

The LORAN-C integration approach can also be extended in terms of passive approach, where the capabilities provided by LORAN-C can be used to improve GNSS performance. In particular, the LORAN propagation properties can be used in environments that lack GNSS signals. A well-known problem of GNSS, in fact, is the poor penetration of L-band signals in signal-blocked environments and, in particular, in urban areas, where the signals are either blocked or reflected by man-made structures. Therefore, in such environments, the GNSS performance, particularly in terms of accuracy, decreases significantly. On the contrary, low-frequency LORAN-C signal propagation is not affected by the built-up areas, since its signal wavelength (equal to 3 km) is considerably longer than the size of the buildings, and, hence, LORAN-C signals can be used to augment GNSS. In this frame, a significant study was undertaken by Delft University that developed, during the 1990s, a European GNSS augmentation system, called Eurofix, based on LORAN-C technology [7, 8]. Eurofix is an integrated navigation system that combines LORAN-C and differential GNSS (DGNSS) (see Chapter 4). The system uses LORAN-C signals to broadcast the differential corrections and integrity information to the GNSS users. The Eurofix system architecture is shown in Figure 9.3.

The correction data are carried over a low-frequency/long-range data channel that is established by an additional three-level time modulation of the LORAN-C pulses. It is worth mentioning that the basic LORAN-C positioning accuracy in not affected by the new applied modulation, and, hence, Eurofix, in addition to the other augmentation systems (see Chapter 4), offers a full navigation back-up system in case of GNSS outage, and, at the same time, allows one to provide GNSS position accuracy of 5m. Eurofix is still under testing and Eurofix messages are currently broadcast on four transmitting stations (Bø, Værlandet, Sylt, and Lessay) of the Northwest European Loran System (NELS), providing DGPS coverage over all of northwestern Europe, as shown in Figure 9.4.

9.2.1.2 Satellite Systems

An interesting aspect in the analysis of GNSS integration with other navigation systems concerns the mutual integration capability of the satellite-based navigation systems themselves. In the global perspective of a pervasive penetration of GNSS services in most fields of the human life, an "active" coexistence of the systems extends their effectiveness in time, space, and performance.

As GPS is the first and, at present, most widespread operational GNSS, it is appropriate to consider the integration issues and features of the integration

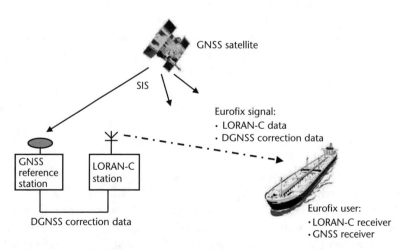

Figure 9.3   Eurofix system architecture.

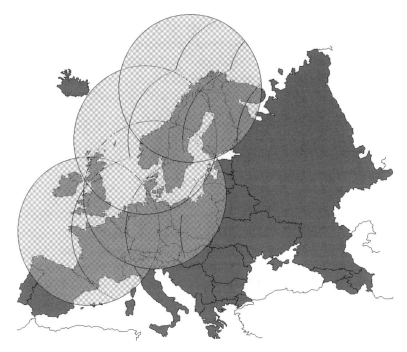

Figure 9.4   Eurofix coverage.

between GPS and the other existing GNSS system, GLONASS (see Chapter 1), as well as the future GALILEO system. Aspects of GALILEO/GLONASS integration were covered in Chapter 5.

As outlined in Chapter 1, GLONASS has many similarities to GPS in terms of constellation parameters and organization as well as signal structure. Unlike GPS, however, each GLONASS satellite transmits its own carrier frequencies in the L1 band (originally 1,602–1,615.5 MHz, to become 1,598.0625–1,604.25 to avoid interference with radio astronomers and operators of LEO satellites, which would otherwise share the same or close ranges) and the L2 band (originally 1,246–1,256.5 MHz, to become 1,242.9375–1,247.75). This shift implies that each pair of GLONASS satellites will be assigned the same L1 and L2 frequencies. The satellite pairs, however, are antipodal; hence they have to be placed on opposite sides of the Earth. The consequence of this is that a user cannot see the two satellites simultaneously. GLONASS codes are the same for all satellites. The GLONASS receiver, hence, uses frequencies other than code to distinguish the satellites. The chipping rates for the two code types (P and C/A) are 5.11 and 0.511 Mbps, respectively, whereas the navigation message is a 50-bps data stream. As highlighted in Chapter 1, in spite of the completion of the satellite constellation in 1996, the number of GLONASS satellites has decreased year by year. The new generation of satellites, called GLONASS-M, is bringing the system back to full operation. The GLONASS-M 11 L satellite, put into orbit in December 2003, has completed its in-orbit tests and has been commissioned; the GLONASS-M 12 L was launched in December 2004 [10].

GPS and GLONASS systems can be effectively integrated to improve geometry and positioning accuracy, particularly under poor satellite visibility, such as in urban areas. It should be mentioned that the GLONASS signal—contrary to the GPS case— has never been affected by either SA or antispoofing. When GPS was adopting SA, the combined use of GPS and GLONASS would have dramatically (about four times) improved the horizontal and vertical error of stand-alone GPS. In the absence of SA, stand-alone and GPS-plus-GLONASS integration performance in terms of accuracy are almost the same [5].

Some issues have to be overcome to achieve a profitable use of the integration [11–16]. GPS and GLONASS systems use different coordinate frames to express the position of their satellites (WGS '84 and PZ-90, respectively): the two systems differ by 20m on the Earth surface. A further issue on GPS/GLONASS integration is the use of different reference times in the two systems: the intersystem time offset changes slowly and reaches several tens of microseconds. The time offset can be determined by considering it as an additional variable in the receiver solution.

Within inter-GNSS integration, a key role is represented by the GPS/GALILEO case. GPS represents the past and present in satellite-based navigation, GALILEO is the future. Interoperability and effective synergy between the two systems is the key to a bright future for satellite navigation. Chapter 5 addressed many issues on the interoperability of the GALILEO system. This is because a new system cannot be conceived and developed without taking into account compatibility and integration as a must. "Compatibility" generally refers to system characteristics of frequency sharing with other GNSS; "interoperability," instead, is referred to as a functional characteristic of a system to be used in combination with others (navigation and non) at the user receiver level.

A picture of the "interoperability" architecture identified for GALILEO and defined during the related *Galilei Project* is provided in Figure 5.16 [17]. Both navigation and nonnavigation systems and interfaces have been studied in terms of coexistence, alternative use, and combined use at the corresponding interoperability level. It was concluded that GALILEO/GPS interoperability has the most to offer from a global GNSS perspective. Application integrators can "play" on a 60-satellite constellation where service applications can be built and offered to the end user. In this context, the optimization of GPS and GALILEO constellation drifts could significantly improve the combined GNSS performance. In addition, the optimum launch sequence for GALILEO satellites should be determined, taking into account the performance benefits in conjunction with GPS.

Significant steps ahead are reported in GPS/GALILEO integration policy documents [18, 19]. In fact, in June 2004 the European Union and the United States concluded an agreement on GALILEO and GPS that allows each system to work alongside the other without interfering with its counterpart signals, thus giving a huge boost to users worldwide [18]. The agreement confirms that GPS and GALILEO services will be fully compatible and interoperable and, thus, renders the joint use of GPS and GALILEO and the manufacturing of equipment much easier and cheaper. The road to the agreement was long and difficult. Although it was soon clear that GALILEO and GPS could complement each other effectively, they are quite dissimilar in important aspects, rendering combining them difficult: This difficulty initially brought some tensions between Europe and the United States. A major dissimilarity between the two systems dwells in their nature: Although GPS serves both civil and military functions, it was originally a military system and is still operated by the armed force of a single nation. It plays a vital role in the security of the United States and European NATO members, in addition to the commercial benefits it brings to both regions. GALILEO, instead, is a civil system, operated by many nations of the European community, with others possibly contributing as well. Although both systems offer open, free-to-air access, GALILEO will provide additional commercial services on a user-pays basis. With real-time integrity, GALILEO will warn users of system failures immediately—as opposed to stand-alone GPS where hours could elapse before a warning is given—and provide legally enforceable service guarantees. The two systems will be obliged by technical and commercial pressures to share common frequency bands and employ compatible codes, timing sources, and geodetic frameworks: user receivers will need to accept both sets of signals [20]. In an interleaved way, GALILEO thus requires U.S. cooperation for its commercial success, while, at the same time, somehow "threatening" U.S. national security and industrial advantage. Nonetheless, GALILEO-plus-GPS is expected to give U.S. civil users dual-frequency positioning capability starting from 2008. Furthermore, since GALILEO and GPS will have separate control segments, vulnerability will be reduced.

A serious question has been posed between GPS and GALILEO about frequencies [21, 22]. GNSS frequencies are very limited; hence GALILEO is obliged to use the same frequency band as GPS. This is also essential to achieve low-cost combined receivers; however, the mutual interference between the two systems was all but a minor issue. This question resulted in extensive research to allow GALILEO signals to be wrapped around GPS signals without negative interference [23–25]. The last frequency-related issue concerned the overlap, which was predicted some time ago, of the GALILEO L1 PRS signal (i.e., the government service) onto the GPS M code, the future encrypted U.S. military signal. This would have been risky: in case of jamming to GALILEO PRS not only would Europe be up in arms, but also the United States would be damaging its own M code. In the current world climate, both Europe and the United States needed to achieve a strong security and noninterference agreement. The agreement was reached [26, 27]. Europe agreed to move the PRS signal away from the M code and outside the declared GPS band. As a result, either the open services or the PRS or M code could be jammed but with a degree of isolation. In addition, Europe and the United States agreed to implement a common signal structure, making it possible for the separate timing and geodetic standards to be interoperable, thus opening the door to a combined system. Furthermore, open trade was also agreed, in which neither Europe nor the United States will mandate the sole use of its own system. Europe and the United States now have a common goal: the best possible GNSS around the world, and this can only benefit users.

As a final remark, it is interesting to report that a single GNSS system can also be "auto-integrated" (i.e., various functions or receivers of a single GNSS can be properly combined for a specific purpose). This is, for instance, the case of the NETNAV system, a GPS/GPS integrated navigation system set up for the Italian Antarctic Research Program (IARP). NETNAV is a complex hardware and software navigation system that is distributed on a local-basis network and can be exploited both in interactive and passive form (thanks to an internal television network). NETNAV uses inputs from four different GPS receivers (two Astech GG24, one Foruno GP500, and one Trimble 4000) and other sensors and distributes the information over the network by using cooperative programs running on two or more PC servers. They watch each other and provide a standard fix using an advanced algorithm; the fix is delivered via the local network and can be received by a large number of PCs located in the various laboratories and on the IARP ITALICA ship [28].

9.2.2 Mechanical Navigation Systems

As already outlined in Chapter 8, the use of GPS for locating or delivering geographically related information requires solutions to the problems arising in particular environments, such as urban canyons, buildings, tunnels, and undercover car parks, where large inaccuracies can be caused by shadowing or even loss of GPS signals, multipath, interference, extended time-to-first-fix (TTFF), and dynamic limitations ascribable to the receiver (e.g., the maximum rate of change in acceleration).

GPS receivers have to provide user position at regular intervals, keeping a continuous knowledge of it between updates. Therefore, these drawbacks have to be solved, for example, by means of *integration of GPS and mechanical sensors technologies*, which should enable performances superior to those of individual technologies. This integration is usually obtained using a *Kalman filter*. Mechanical sensors here refer to inertial sensors, such as accelerometers, gyroscopes, odometers, speedometers, altimeters, dopplometers, and so on. As indicated in Chapter 8, these devices can be grouped under the INS banner. In particular, when the mechanical system consists of an odometer and a vibration gyroscope, the system is referred to as DR. The GPS/DR system is widely used in automatic vehicle location (AVL) [29].

The realization of a GPS/INS device (or, equivalently, GPSI device, with "I" meaning inertial) comes from considering various aspects (i.e., its physical characteristics such as size, weight, or power consumption), required performance (application-related and/or environment-related), and costs (both recurring and nonrecurring) [5].

From the physical characteristics point of view, the reduction of size and weight can be achieved through the use of *microelectromechanical systems* (MEMS), replacing the more cumbersome, conventional inertial sensors.

The performance achievable with integration must take into account not only periods of GPS signal reception outages but also the output rate of position updates, which can be higher than that obtained by GPS alone. In addition, the synergy between satellite navigation and inertial systems can mitigate the random component of errors in the GPS solution.

Cost of integration is a factor influenced by the chosen type of inertial sensor.

Inertial navigation systems benefit from integration with GPS in two main ways [30]:

- The inertial navigation errors are bounded by the accuracy of the GPS solution.
- The GPS receiver can be used to calibrate the inertial sensors.

Inertial sensors are usually chosen for integration with GPS receivers because they are passive, autonomous, and available; in addition, they are immune to GPS causes of outage. Their adoption can enrich the information provided by a navigational system with attitude (rotational) data. Furthermore, the higher output rate of inertial sensors can furnish position data between two successive GPS updates. Besides providing a more complete set of navigation information, the introduction of INSs in satellite positioning technology allows the reduction of noise in GPS navigation solutions. Stand-alone GPS receiver processors usually propagate the previous navigation solution to the current measurement epoch in a linear filtering algorithm. However, GPS does not measure acceleration, so that the propagated solution is sensitive to errors in previous acceleration estimates or to changes in acceleration during the propagation interval. On the contrary, inertial systems can sense acceleration in the interval between GPS updates. Implementing a well-tuned Kalman filter and using GPS measurements to obtain error estimates in the INS output can reduce the effect of additive noise on any update because the filter smoothes the INS output update over many GPS measurements.

Finally, increased tolerance to dynamics and interference can be achieved through feeding back the INS velocity solution to the GPS receiver [5].

9.2.2.1 Inertial Systems Review

Before dealing with GPS/INS integration, a brief review of traditional inertial system technology is given here. Future trends are summarized at the end of this section.

Inertial sensors and systems standards are developed by the Gyro and Accelerometer Panel of the Institute of Electrical and Electronics Engineers Aerospace and Electronics Systems Society (IEEE/AESS) and published by the Standards Association of the IEEE.

Different types of inertial systems defined over the 40-year life (to date) of the Panel Organization include [31]:

- *Inertial Sensor Assembly (ISA):* It consists of multiple inertial sensors, such as gyroscopes and/or accelerometers, having fixed orientations relative to each other.
- *Attitude and Heading Reference System:* It estimates the body angles of a vehicle in the local level coordinate system, using accelerometers to determine the level and gyroscopes to track the north direction without external reference.
- *Inertial Reference Unit (IRU):* It provides measurements of inertial angular motion in three dimensions without external reference, employing, for example, gyroscopes.
- *Inertial Measurement Unit:* it measures linear and angular motion in three dimensions without external reference.
- *Inertial Navigation System:* Its outputs are the estimates of a vehicle position, attitude, and velocity as a function of time in a navigation frame using the outputs of an IMU, a reference clock, and a model of the gravitational field. INS can be of two types: *strapdown INS* (Figure 9.5) [31], whose ISA is fixed relative to the vehicle body, and *gimbaled INS* (Figure 9.6) [31], which can be inertially oriented (i.e., the ISA is fixed relative to an inertial reference, such as

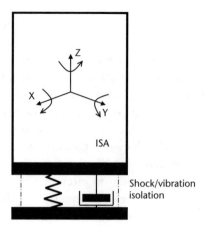

Figure 9.5   Strapdown INS.

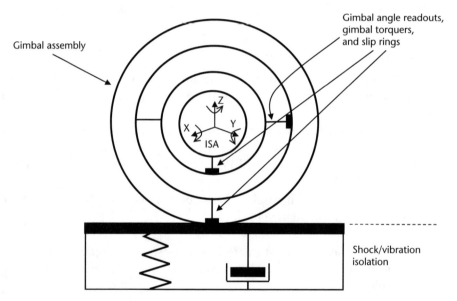

Figure 9.6   Gimbaled INS.

that specified by distant star directions), local-leveled (i.e., the ISA is aligned with a local level coordinate system), indexed (indexing is the discrete rotation of the ISA about one or more axes to mitigate navigation errors), carouseled (carouseling is the continuous rotation of the ISA about one or more axes to reduce navigation errors), rate biased (i.e., the ISA has a large rotation rate about an axis due to gimbal or vehicle rotation), or thrust-following (i.e., the ISA has one axis along the thrust vector of the vehicle).

Calibration and initialization are fundamental to allow an inertial system to operate. Both a strapdown and gimbaled INS can be calibrated by comparing INS output and GPS output; also, the initialization information, composed of position and/or velocity of the system, can be obtained by means of GPS. In addition, another important issue is the synchronization of the INS clock with the GPS one. The Kalman filter is usually used to estimate the system and sensor model parameters required for calibration [31].

The following is a list of conventional inertial sensors, highlighting their functionalities and employment in the field of vehicle navigation:

- A *gyroscope (or gyro)* measures angular rotation with respect to inertial space about its input axis(es), using, for example, the angular momentum of a spinning rotor or the Coriolis effect on a vibrating mass.

- A *linear accelerometer* measures the component of translational acceleration minus the component of gravitational acceleration along its input axis(es), sensing the motion of a proof mass relative to the case or measuring the force or torque necessary to carry back the proof mass to a null position relative to the case.

In particular, it is worth mentioning the following specific types of inertial sensors: the *ring laser gyroscope* (RLG), invented in the 1960s and replacing electromechanical instruments by the late 1980s and early 1990s because of its great applicability in strapdown cases; the *dynamically tuned gyroscope* (DTG), also conceived in the 1960s with the aim of offering two axes of rate information in one sensor; the *hemispherical resonant gyroscope* (HRG), which is a high-performance vibrating shell gyro used in spacecraft IRUs [32]; and *interferometric fiber-optic gyroscope* (IFOG), which measures angular rotation from the interference pattern between counterpropagating laser beams through an optical fiber. In regard to accelerometers, we can cite *mechanical pendulous force-rebalance accelerometers*, *vibrating beam accelerometers* (VBAs), and *gravimeters* (input-axis-vertical limited-range accelerometers) [31, 33, 34].

The performance factors useful to compare inertial sensors can be the *scale-factor stability*, which indicates the ability of the sensor to reproduce the sensed rate or acceleration, and the *bias stability*, where the bias is the measurement error independent of rate or acceleration of the system incorporating the sensor.

As far as gyroscopes are concerned, mechanical (i.e., spinning mass) and ring laser are those offering high performance but having high costs and being available in limited quantity; their fields of application include tactical navigation, autonomous submarine navigation, and air/land/sea navigation surveying. DTGs, which share some applications with RLGs, offer medium-level performance. IFOGs, Coriolis-based sensors and rate and integrating gyroscopes, offer lower performance but are less expensive and available in high quantity. They are employed in AHRS torpedoes, tactical missile midcourse guidance, flight control, smart munitions guidance, and robotics [33] (Figure 9.7).

In the case of accelerometers, mechanical floated instruments offer the best performances and find application in self-alignment of strategic missiles and in the stellar-aided strategic missile field. Next in terms of performance are the mechanical pendulous rebalance accelerometers used in autonomous submarine navigation, cruise missile navigation, land and aircraft navigation, and stellar-aided reentry. Quartz resonator accelerometers, which are used in lower-grade tactical and commercial applications, offer the lowest level of performance (Figure 9.8) [33, 34].

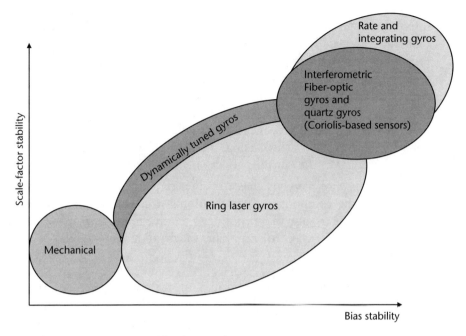

Figure 9.7   Current gyroscope technology performance.

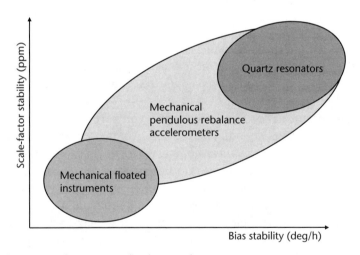

Figure 9.8   Current accelerometer technology performance.

9.2.2.2   GPS/INS Integration Architectures

Integration typologies are very different from each other, depending on mission requirements and the budget available to satisfy these requirements.

All possible realization schemes can be traced back to three system architectures:

- *Uncoupled mode* (Figure 9.9) [5];
- *Loosely coupled mode* (Figure 9.10) [5];

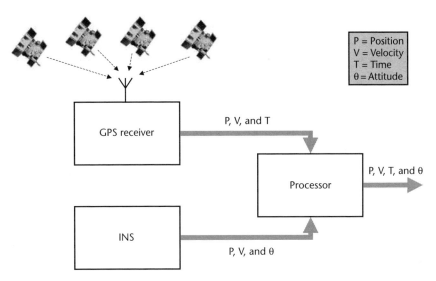

Figure 9.9 Uncoupled GPS/INS integration.

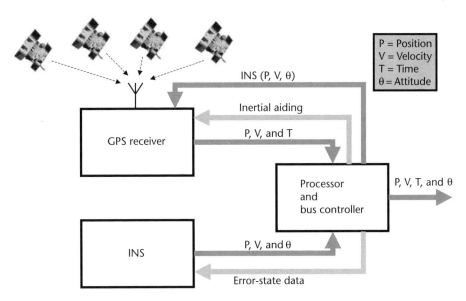

Figure 9.10 Loosely coupled GPS/INS integration.

• *Tightly coupled mode* (Figure 9.11) [5].

In the *uncoupled* mode, the GPS user equipment and inertial system provide independent solutions, which are integrated by a processor performing either a simple selection between the two navigation data type or a multimode Kalman filtering; the data travels through unidirectional busses to reach the processor. This type of integration is the most straightforward and potentially cheapest solution. In addition, if one of the two subsystems fails, the other still operating continues supplying navigation information with its own level of accuracy [5].

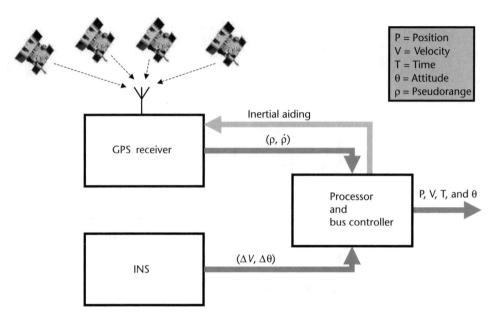

Figure 9.11   Tightly coupled GPS/INS integration.

The *loosely coupled* (or integrated) configuration was first adopted in the early 1980s [30]; unlike the uncoupled mode, it presents many data paths, the most important of which is feeding back the system navigation solution to the GPS receiver. As already mentioned, the GPS user equipment cannot directly sense acceleration, so it has to use noisy acceleration estimates based on recent velocity measurements for the propagation of the previous navigation solution to the epoch of the current tracking loop outputs. Performing this propagation using the navigation solution determined by the processor is of great improvement, because this solution includes INS measurements accounting for acceleration; if the filter memory capacity is extended, the averaging operation of the filter can be more effective with regards to the GPS measurement noise. Another data bus provides inertial aiding to the tracking loops of the GPS receiver, reducing the vehicle dynamics that these loops have to track. In principle, INS should send the aiding directly to the GPS user equipment, but this approach is discarded to avoid implementing custom interfaces between the two navigational subsystems; therefore, the inertial aiding is applied to the GPS user equipment by the processor, which has just done the conversion of data coming from INS from inertial coordinates to GPS coordinates. Also the INS has its own feedback path to receive error-state data, hence being able to correct its position and velocity and realign its platform [5].

In the *tightly coupled* mode, the GPS receiver and inertial subsystem do not perform filtering operations, but they only transmit, respectively, GPS code and carrier measurements and acceleration and angular rate to the processor. Concerning feedback buses, this configuration presents only one path from the processor to the GPS receiver to provide the latter with velocity data aiding.

Tightly coupled integration is often associated with embedded GPS receivers, in which the inertial sensors are physically and electrically integrated with them, even though "tightly coupled" does not mean "embedded" [5].

9.2.2.3   GPS/INS Integration Algorithms

Each one of these integration schemes can be associated with any of the following algorithms [5]:

- *Selection* (with or without inertial system reset): the processor chooses the GPS navigation solution when this one is accurate, reverting to an INS solution whenever a higher output rate is needed or during GPS outages.

- *Filtering*: the aim of filtering algorithms is to give an estimate of the states of a system described by differential equations, using measurements of states-related quantities. When the current estimate of the state is given by a weighted sum of the measurements and the previous state value, the filters are referred to as *linear* (i.e., linear filters); the weight associated with measurements is referred to as *filter gain*. In the case of GPS/INS integration, the measurements used as input to the filter are the differences between GPS position and INS position and the differences between GPS velocity and INS velocity. Two different types of filter can be identified according to the values of the filter gain:

  - *Fixed-gain filtering*: gain values are determined before the filter operates, hence dismissing the processor from further computational tasks and memory storage. This filtering approach is viable when the system presents bounded state dynamics and state dynamics uncertainty and negligible variation of measurement noise.

  - *Time-varying gain*: gains are updated at every instant of filter computation; the most used discrete-data linear filter with time-varying gain is the one conceived by R. E. Kalman in 1960.

Kalman Filter

The Kalman filter is a recursive algorithm, which determines the optimal state estimate at a given instant, using only the measurements at that instant and the optimal state estimate obtained at the previous instant.

To appreciate recent developments in GPS/INS integration algorithms based on the use of a Kalman filter it is necessary to review its basic equations.

A discrete-time nonlinear system, affected by system error and measurement error, can be generally described with the following equations:

$$\underline{x}(k+1) = f\left[\underline{x}(k), \underline{u}(k), \underline{w}(k)\right]$$
$$\underline{y}(k) = h\left[\underline{x}(k), \underline{v}(k)\right]$$

(9.3)

where:

- $\underline{x}(k)$ is the state vector at the kth instant of time.
- $\underline{u}(k)$ is the input vector to the system at the kth instant of time.
- $\underline{y}(k)$ is the output vector of the system at the kth instant of time.
- $\underline{w}(k)$ and $\underline{v}(k)$ are the system error vector and measurement error vector, respectively.

- $f(.)$ is the function describing the system evolution.
- $h(.)$ is the function describing the measurement process.

Equation (9.3) fitted for linear systems assumes the following form:

$$\underline{x}(k+1) = A(k)\underline{x}(k) + B(k)\underline{u}(k) + \tilde{F}(k)\underline{N}'_k$$
$$\underline{y}(k) = C(k)\underline{x}(k) + D(k)\underline{u}(k) + \tilde{G}(k)\underline{N}''_k \tag{9.4}$$

where the stochastic inputs to the system, $\tilde{F}(k)\underline{N}'_k$ (state noise vector) and $\tilde{G}(k)\underline{N}''_k$ (measurement noise vector), are shown explicitly; due to their nature, they are known only through their statistical characteristics (i.e., probability density function or mean value and covariance [35]). The aim of the Kalman filter is to determine the best estimate of the state $\underline{x}(k)$ given the set of control input $\underline{u}(0)$, $\underline{u}(1)$, ..., $\underline{u}(k-1)$, the set of measurements $\underline{y}(0)$, $\underline{y}(1)$, ..., $\underline{y}(k-1)$, $\underline{y}(k)$, and the initial conditions. The definition "best estimate," in the case of Kalman filter, means the minimization of the mean-square error between the state vector estimate and the true state vector (i.e., the minimization of the function):

$$J(\underline{\tilde{x}}(k)) = E\left\{\|\underline{\tilde{x}}(k) - \underline{x}(k)\|^2\right\} \tag{9.5}$$

where $E(.)$ indicates the expectation function.

To improve readability of (9.4) it can be defined:

$$\underline{N}_k = \left[\frac{\underline{N}'_k}{\underline{N}''_k}\right] \tag{9.6}$$

and

$$F(k) = \left[\tilde{F}(k) \quad 0\right] \tag{9.7}$$

$$G(k) = \left[0 \quad \tilde{G}(k)\right] \tag{9.8}$$

so (9.4) becomes:

$$\underline{x}(k+1) = A(k)\underline{x}(k) + B(k)\underline{u}(k) + F(k)\underline{N}_k$$
$$\underline{y}(k) = C(k)\underline{x}(k) + D(k)\underline{u}(k) + G(k)\underline{N}_k \tag{9.9}$$

where $\left\{\underline{N}_k\right\}$ is a white Gaussian sequence with zero mean:

$$E\left\{\underline{N}_k\right\} = \underline{0} \tag{9.10}$$

$$E\left\{\underline{N}_k \underline{N}_j^T\right\} = \delta_{k,j} I, \quad \forall i, j \tag{9.11}$$

also the initial state $\underline{x}(0)$ is supposed to be a casual Gaussian vector with zero mean, uncorrelated to the noise $\overline{N_k}, \forall k$.

The *best state vector estimate* determined by means of Kalman filter is given by the equation:

$$\hat{\underline{x}}(k) = A(k-1)\hat{\underline{x}}(k-1) + K(k)\lfloor \underline{y}(k) - C(k)A(k-1)\hat{\underline{x}}(k-1) \rfloor \tag{9.12}$$

where $K(k)$ is the *Kalman gain matrix*, which can be written as follows:

$$K(k) = P_p(k)C^T(k)\left[C(k)P_p(k)C^T(k) + G(k)G^T(k) \right]^{-1} \tag{9.13}$$

with $P_p(k)$ defined as the covariance of the prediction error vector $\underline{e_p}(k)$, which, in turn, is given by:

$$e_p(k) = \underline{x}(k) - A(k-1)\hat{\underline{x}}(k-1) \tag{9.14}$$

$$P_p(k) = E\left\{ \underline{e_p}(k)\underline{e_p^T}(k) \right\} = A(k-1)P(k-1)A^T(k-1) + F(k-1)F^T(k-1) \tag{9.15}$$

The matrix $P(k)$ is the covariance of the estimate error vector and is defined as:

$$P(k) = E\left\{ \underline{e}(k)\underline{e}^T(k) \right\} = \left[I - K(k)C(k) \right]P_p(k)\left[I - K(k)C(k) \right]^T + K(k)G(k)G^T(k)K^T(k) \tag{9.16}$$

where $\underline{e}(k) = \underline{x}(k) - \hat{\underline{x}}(k)$ is the estimate error vector and $E\{\underline{e}(k)\} = \underline{0}$, given that $E\{\underline{x}(k)\} = E\{\hat{\underline{x}}(k)\}$ [35].

Equations (9.13), (9.15), and (9.16), which iteratively give the sequence of the covariance matrices $P(k)$, are called *Riccati formulas*; the *Riccati equation* expresses $P(k)$ as a function of $P(k-1)$.

A Kalman filter schematic representation is shown in Figure 9.12 [35].

A simple GPS/INS integration example providing Kalman filter computations is presented in [30], where the subsystems are a single gyroscope and a single-channel GPS receiver whose antenna shares the same origin with the INS. As the state vector $\underline{x}$, the authors chose an error state one, including position and velocity errors, δx and $\delta \dot{x}$, respectively, and the GPS receiver clock bias and clock drift, δt and $\delta \dot{t}$, respectively:

$$\underline{x} = \begin{bmatrix} \delta x \\ \delta \dot{x} \\ \delta t \\ \delta \dot{t} \end{bmatrix} \tag{9.17}$$

The transition matrix and the measurement matrix indicated in [30] correspond to matrix A and B, respectively, according to our notation.

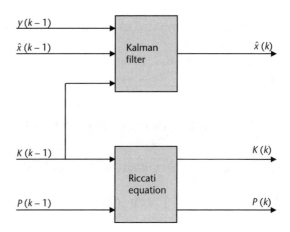

Figure 9.12   Schematic representation of the Kalman filter operations.

The steps to obtain an estimate of the error state vector (9.17) are then performed.

In the case of multisensor systems, fault-tolerance can be achieved using the method of the *federated Kalman filter*, where filtering tasks are divided in two stages: there is a master filter and various local filters corresponding to the different inertial sensors. These local filters process their data providing the best local estimates; then, the master filter integrates the estimates coming from the local filters to obtain the best global estimate. If a single Kalman filter had been used, its computational burden would have been very high. Furthermore, if one sensor had sent a faulty datum, the result achieved by the single Kalman filter would have been wrong [36].

The Kalman filter is based on a dynamic model of errors development over time and on a stochastic model of the noise affecting each sensor. In the presence of time-varying characteristics of the errors, using "a priori" statistics makes the stochastic model of the errors limited; so, an improvement of the stochastic information is necessary, leading to the development of the so-called *adaptive Kalman filtering* to indicate that the stochastic information is modified on-line. In [37] three algorithms of such a type are presented: the *covariance scaling/process noise scaling*, the *adaptive Kalman filter* (AKF), and the *multiple model adaptive estimation* (MMAE). Generally, an adaptive Kalman filter has to identify the correct stochastic properties without any "a priori" statistics. In addition, it must be able to act efficiently in real time, timely responding to new observations and modeling any correlation between observations or states.

The *covariance scaling filter* [37] is very simple, consisting of multiplying the covariance of the prediction error vector $P_p(k)$, given by (9.15), by a scale factor $S(k) \geq 1$, obtaining the modified covariance of the prediction error vector $\tilde{P}_p(k)$

$$\tilde{P}_p(k) = S(k)\left\{A(k-1)P(k-1)A^T(k-1) + F(k-1)F^T(k-1)\right\} \qquad (9.18)$$

In the case of $S(k) > 1$, more weight is put on new measurements.

If the scale factor is applied only to the process noise factor, $F(k-1)F^T(k-1)$, the filter is referred to as a process noise scaling filter:

$$\tilde{P}_p(k) = A(k-1)P(k-1)A^T(k-1) + S(k)F(k-1)F^T(k-1) \qquad (9.19)$$

The *MMAE algorithm* exploits multiple Kalman filters running simultaneously and each using a different stochastic model [37]. Then, the correct filter to use can be identified according to different algorithms or, alternatively, the state vector estimates produced by the filters can be combined together, achieving the optimal state vector estimate

$$\underline{\hat{x}}(k) = \sum_{i=1}^{N} p_i(k)\underline{\hat{x}}_i(k) \qquad (9.20)$$

where $\hat{x}_i(k)$ is the state vector estimate determined by the ith filter of the bank, N is the number of filters, and $P_i(k)$ is the probability that the ith model is correct.

AKF is the name of a *particular* algorithm in the set of generic adaptive Kalman filters [37]; it is an extension of the conventional Kalman filter. Indicating with $Q(k)$ the process noise matrix

$$Q(k) = F(k-1)F^T(k-1) \qquad (9.21)$$

this algorithm determines the adaptive estimate of the process noise matrix, $\hat{Q}(k)$, using the following expression

$$\hat{Q}(k) = \hat{C}_{\Delta x(k)} + P(k) - A(k-1)P(k-1)A^T(k-1) \qquad (9.22)$$

in which $\hat{C}_{\Delta x(k)}$ is the state correction covariance matrix.

9.2.2.4   Inertial Sensor Technology Trends

Inertial sensor technology is moving toward improving *fiber-optic gyroscopes* (FOG) and the development of *MEMS gyroscopes and accelerometers* [33].

FOGs were invented in the 1960s as well as *ring laser gyroscopes,* but their development had been forced to track that of fiber-optic technology, resulting in slower diffusion in possible fields of application. FOGs principle of operation is based on the Sagnac effect [34], involving the use of a fiber-optic sensing coil of a length variable from meters to a kilometer, an integrated optics chip, a broadband light source, and a photodetector. A reduction of size and cost can be obtained substituting this configuration with quantum well technology. Furthermore, FOGs should replace RLGs because of their lower costs, especially in the lower-performance tactical and commercial applications.

Micromechanical gyros are usually realized as electronically driven resonators, exploiting the Coriolis force generated when an angular rate is applied to a translating mass (Figure 9.13) [38]: in the case of a micromechanical tuning fork gyro, the application of an angular rate to the axis of the fork makes its tines subjected to the Coriolis force, which causes torsional forces about the sensor axis. These forces, proportional to the applied angular rate, provoke displacements measurable capacitively in a silicon instrument or piezoelectrically in a quartz one. This type of gyros

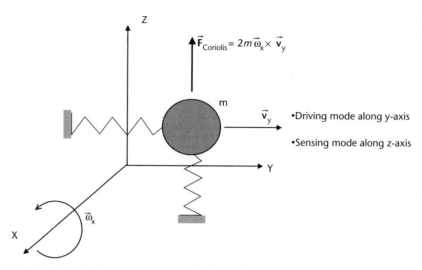

Figure 9.13   Physics of a vibratory gyroscope.

is suitable for very high acceleration applications, because of their small size and the strength of silicon [33].

Micromechanical accelerometers are generally composed of a proof mass suspended by beams anchored to a fixed frame; the mathematical model of an accelerometer is a second-order mass-damper-spring (Figure 9.14) [38]. When an external acceleration causes a displacement of the support frame relative to the proof mass, the latter produces a variation of the stress of the spring; to gain a measure of the external acceleration, both the relative displacement and the spring stress can be adopted. The transduction mechanisms exploited in micromechanical accelerometers cover a broad spectrum: piezoresistive, capacitive, tunneling, resonant, thermal, electromagnetic, and piezoelectric. Silicon accelerometers are used in many applications, ranging from automotive air bags and ride control to competent munitions and autonomous vehicles. Quartz resonant accelerometers are purchased for tactical and commercial (e.g., factory automation) applications [33, 38, 39].

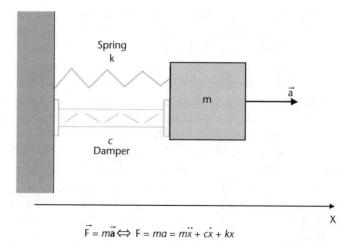

$$\vec{F} = m\vec{a} \Longleftrightarrow F = ma = m\ddot{x} + c\dot{x} + kx$$

Figure 9.14   Physics of a linear accelerometer.

Optical MEMS sensors, also indicated as *micro-optical-electromechanical systems* (MOEMSs), are the frontier of MEMS technology, because they will be available when low-loss waveguides and very narrowband light sources are developed. MOEMSs will be true solid-state devices [33].

9.3 Integration with Communications Systems

9.3.1 2G/3G Mobile Networks

In the last few years, the FCC E-911 mandate has been the driving force for integrating GPS receivers into wireless handsets. The rule requires that, as of October 1, 2001, public safety answering point (PSAP) attendants have to be able to locate a 911 mobile caller, with the accuracy levels reported by Table 9.3 [40, 41].

In Europe, the European Commission has taken similar initiative for E-112 calls.

The date of October 1, 2001, had been fixed by the U.S. FCC as the beginning of Phase II of the wireless E-911 program; this phase is scheduled to be completed by December 31, 2005, and it follows Phase 0 and Phase I. The Phase 0 requirement was only that wireless 911 calls had to be transmitted to a PSAP regardless of whether the caller was a wireless subscriber or not. The Phase I requirement was that a local PSAP had to know the phone number of a wireless 911 caller and the location of the antenna that received the call [42, 43].

The FCC has been endeavoring to promote wireless E-911. In 2001, it commissioned the "Hatfield Report" [42] to study E-911 technical and operational deployment; in 2002, the FCC undertook a rulemaking to evaluate the possibility to extend E-911 rules to other communications services. Finally, in 2003, the FCC endorsed the Wireless E-911 Coordination Initiative to hasten E-911 operation through the exchange of experiences among stakeholders and the definition of possible development strategies. The emphasis had been put on many issues comprising PSAP upgrades, wireless carrier implementation and prioritization, local exchange carriers (LEC) concerns, and problems faced by rural carriers [42, 43].

The location requirements of Table 9.3 can be satisfied outside urban areas only through the means of satellite navigation; hence, the interest in positioning techniques based on the use of a GNSS combined with the cellular network infrastructure is very strong.

Actually, we can say that GNSS and cellular networks can be considered complementary technologies: GPS or GALILEO provide the means for complying with the U.S. FCC E-911 mandate and even achieving more stringent requirements, whereas cellular networks also help a GNSS to have good performance in environments hostile to the propagation of satellite signals such as urban canyons, building interiors, dense foliage, tunnels, underground parking lots, and so on.

Table 9.3 Accuracy Required for Locating Mobile Phones

Solutions	67% of Calls (m)	95% of Calls (m)
Handset-based	50	150
Network-based	100	300

Mobile phone positioning methods for 2G/3G systems are *cell-ID*, *time of arrival*, *time difference of arrival* (TDOA), *enhanced observed time difference*, *advanced forward link trilateration* (A-FLT), *angle of arrival* (AOA), and *assisted-GPS* [44–47].

Besides compliance with the U.S. FCC E-911 mandate regarding safety-of-life concerns, the purpose of locating a mobile phone and hence its user also comes from the growing need for service providers to offer their subscribers location-based services, such as localized information (local hotels, restaurants, bars, banks) and commercial information (advertising, special promotions, and so on).

In addition, the knowledge of the location of a mobile phone can be exploited to increase the capacity of a CDMA network [48] or plan new handoff strategies to minimize the possibility for interruption, hence increasing the quality of service offered by the network [49, 50].

9.3.1.1 Assisted-GPS

A-GPS was mentioned in Chapter 8, which described its application in indoor environments. Here, its architectural elements and operational characteristics are explained to highlight the supporting role of cellular networks.

A-GPS implementation arises from the possibility of exploiting a bidirectional wireless link to convey data between a mobile station and the cellular network when the mobile station is equipped with a partial or full GPS receiver, where "partial" indicates that the network performs the location calculations (this situation is defined as "user equipment-assisted GPS" later in this section). According to terminology used in the Universal Terrestrial Radio Access Network (UTRAN) and in the GSM EDGE Radio Access Network (GERAN) [45], a mobile station or handset is indicated by UE, whereas a base station by node B. In addition, when the UE position is calculated at the network, it is a UE-assisted solution, whereas when the UE position is determined at the mobile phone, it is a UE-based solution.

As already outlined, A-GPS uses a GPS reference network, formed by many *A-GPS servers* or *location servers*, or a WADPGS network, connected to the cellular infrastructure and consisting of GPS receivers that can detect GPS signals continuously and monitor the satellite constellation in real time (Figure 9.15).

The reference network has several tasks, such as providing approximate UE position or node B locations, satellite parameters, Doppler data, and others. The assistance information is transmitted from the GPS reference network to the UE, achieving the following benefits [51]:

- Shrinking the start-up (i.e., the TTFF from 30 seconds to a few seconds);
- Enhancing the sensitivity of a GPS receiver, allowing it to determine position even in a difficult environment;
- Improving the position accuracy to the level obtained with DGPS.

In conventional GPS receivers, the phase of acquisition of GPS satellite signals is practically a search over the whole possible frequency and code-delay space: the satellite motion introduces a Doppler shift $\pm \Delta f_{D,sat}$, to which a further Doppler shift, $\pm \Delta f_{D,rec}$, due to the eventual receiver motion, has to be added. In addition, the

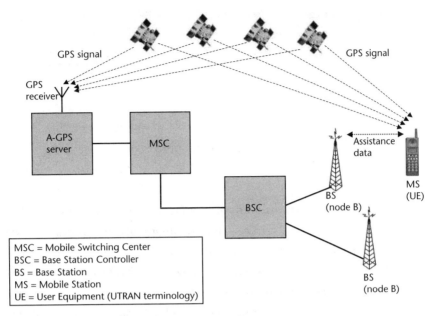

Figure 9.15 Structure of A-GPS.

Doppler shift on the PRN code of the GPS receiver, $\pm\Delta f_{D,PRN}$, should also have been considered, but it is much smaller than the other two because of the low frequency of the PRN code. So, the receiver has to search a space of dimensions $\left(\pm\Delta f_{D,sat} \pm\Delta f_{D,rec}\pm\Delta f_{D,PRN}\right) \times 1{,}023$ to acquire one GPS satellite signal [Figure 9.16(a)]. Therefore, to achieve a first-fix, such a space has to be explored at least four times, resulting in a relatively long TTFF.

If the GPS receiver knew the approximate Doppler shift (as the algebraic sum of all mentioned components), the TTFF would be shorter [Figure 9.16(b)]. A major task of an A-GPS server is to provide the GPS receiver with this approximate Doppler shift, from the knowledge of the coarse position of the UE, obtained from the mobile switching center (MSC), connected to the base station controller (BSC) relative to the base station (BS) cell/sector where the UE is located (Figure 9.15).

Assistance information enables the receiver tracking loop bandwidth to be tightened, increasing the receiver sensitivity, enabling it to send signals with low strength. Furthermore, the reduced bandwidth loop filter takes out more noise, improving the position accuracy.

Another improvement to the detection of GPS signals by the GPS receiver is made available by the *sensitivity assistance*, also known as *modulation wipe-off*. It is a message with predicted navigation bits, which are guessed to modulate the GPS signal transmitted by a given satellite at given times [52].

Other assistance data can be code-delay estimates [53], real-time integrity, DGPS corrections, satellite almanac, ionospheric delay, and UTC offset [45].

An A-GPS UE can run either in one or both of the following modes [45, 51]:

• *UE-assisted (or MS-assisted) GPS;*
• *UE-based (or MS-based) GPS.*

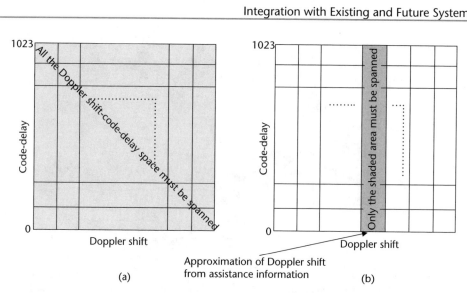

Figure 9.16   GPS signal search space (a) without and (b) with assistance information.

In the *UE-assisted* solution, the UE acquires GPS signals, makes measurements by correlating the locally generated PRN codes with the received GPS signals, and determines time-stamped pseudoranges that are transmitted to the A-GPS server, which performs the calculation of the UE position. In particular, the exchanged information elements (IE) are:

- From the UE to A-GPS server:
  - Location request.
  - Coarse position of the UE.
  - Time-stamped GPS pseudoranges.
- From the A-GPS server to the UE:
  - Visible satellite list;
  - Satellite-in-view signal Doppler and code phase or, alternatively, approximate handset position and ephemeris;
  - GPS reference time.

The location server can also apply differential corrections to the pseudorange received from the UE or to the position value it has determined. The UE has to include an antenna, an RF section and a digital processor, since it acts as a data collection front-end for the A-GPS server.

On the contrary, in the *UE-based* solution, the UE has to incorporate a fully functional GPS receiver (i.e., the equipment necessary not only to acquire GPS signals but also to compute the UE position). The transferred IEs are part or all of:

- From the UE to A-GPS:
  - Location request;
  - Coarse position of the UE;
  - Calculated position of the UE.
- From the A-GPS server to the UE:

- Visible satellite list;
- Satellite-in-view signal Doppler and code phase;
- Precise satellite orbital elements (ephemeris), which are valid for 2–4 hours and extendable to the entire period of satellite visibility (12 hours);
- DGPS corrections;
- GPS reference time;
- Real-time integrity (failed/failing satellite information).

The UE-based GPS has relatively short uplink IEs, whereas downlink assistance IEs are relatively long; the opposite is true for UE-assisted GPS.

A UE-based solution can be chosen for tracking or navigation applications because assistance IEs are valid for 2–4 hours or up to 12 hours at the UE if the ephemeris life extension feature is used. A UE-assisted solution can be adopted for tracking or navigation applications, only taking more signaling into account, because of the much shorter time validity of UE-assisted GPS assistance IE compared with the time validity of UE-based GPS assistance IE.

The UE-assisted solution saves computing power and memory at the UE, whereas the UE-based solution can also be used as a stand-alone GPS receiver [45].

Hybrid Solutions

A-GPS and the cited 2G/3G positioning methods have different levels of accuracy and coverage in different environments. Therefore, more robust locating techniques result from hybrid approaches (i.e., from the combination of A-GPS and mobile phone-based positioning), since they can compensate each other. In dense urban environments, GPS signals undergo shadowing or blockages, whereas these environments are the best for 2G/3G positioning methods, since there is the highest density of BSs. On the contrary, open-sky spaces give prominence to A-GPS, because GPS is at the best end of its level of coverage, whereas the low density of BSs does not allow one to achieve high accuracy with network-based solutions [51, 52].

Examples of hybridization of A-GPS and cellular positioning methods are:

- *A-GPS/Cell-ID* [54]: Cell-ID is the worst cellular positioning method in terms of accuracy, because the MS position is identified with the serving BS position; however, cell-ID positioning is available in all cellular networks.
- *A-GPS/TDOA-TOA* [44]: TOA allows one to determine the distance between the MS and BS by measuring the time interval a radio wave takes to propagate between them (also used in GPS pseudorange measurements, see Chapter 2); this technique requires the synchronization of the MS and BS clocks [51]. TDOA is a hyperbolic locating technique, based on the difference between the time intervals of propagation of a signal from an MS to two BSs; this time difference is converted to a distance difference of the MS to the two BSs, since there is direct proportionality between traveled distance and time interval taken by the signal to go from the MS to each BS. Making this distance difference equal to a constant, a hyperbola is defined. Using three pairs of BSs, the MS position is univocally determined in the two-dimensional case [44, 51] by the intersection of the three hyperbolas having the BSs as foci. The accuracy of

the TDOA technique depends on the relative BS geometric locations and the synchronization of the MS and BS clocks.

- *A-GPS/E-OTD* [55]: E-OTD [44, 47] is a TDOA location technology used in GSM networks (asynchronous networks), in which the observed time difference (OTD) feature already exists.
- *A-GPS/A-FLT* [56, 57]: A-FLT [44] is a TDOA implementation used in CDMA networks (synchronous networks), which are time-synchronized, making easier time difference measurements (i.e., phase-delay measurements between CDMA pilot signal pairs).

A-GPS/cell-ID is a straightforward integration that offers great roaming advantages and can be employed in high population networks [54].

In [44], the motivation for the implementation of a hybrid positioning technique is considered in the particular case of integrating A-GPS and *TDOA* or *TOA*. In addition, a situation in which A-GPS may be a valid support for TDOA/TOA is presented. For example, when moving along a straight highway, TDOA has difficulty providing position solutions, because the "heard" BSs most probably lie along the highway, resulting in an unfavorable geometric configuration. The integration of A-GPS and TDOA/TOA overcomes such adverse cases. The position solution can be achieved, in a 2D space, with a configuration of a minimum of three reference points (i.e., one satellite and two BSs, one BS and two satellites, and so on).

The same argumentations are reported in [55], where the delivery of nonstandardized assistance information is presented, in particular, for the case of the *EMILY Project* [58, 59]. It is a hybrid MS-based *E-OTD/A-GPS* location system, where, besides the conventional assistance data, the real time difference between the base station serving the MS and the satellite, RTD_{BS-SAT}, is used (Figure 9.17).

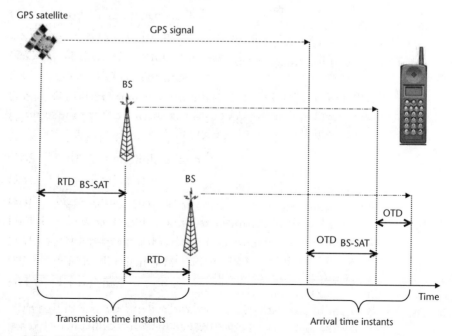

Figure 9.17 EMILY location determination procedure.

Technical Solutions for Integration

The integration of GNSS and cellular positioning techniques requires the development of new technologies for handset design and assembly. Since these aspects involve manufacturing processes and economical issues, examples of devices already developed cannot be avoided, hence the references to various producers and carriers. A fuller list consists of these and many more to come.

The handsets made to date embed GPS receivers and cellular phones of all digital generations. Therefore, the assistance data provided to the GPS receiver can be delivered using SMS or GPRS transmission channels, which can offer more advantages than GSM transmission, because GPRS provides better means for the routing of data packets.

As the influence of satellite-based navigation increases in many aspects of our life (see Chapter 8), research has to look for ways to share functional components in integrated GNSS/mobile phones to realize smaller, cheaper, and less power-consuming devices.

In regard to 3G cellular handsets, the possibility of using some parts of a mobile phone receiver for both satellite and wireless signals detection has been investigated [60]. In addition, the signal structures of GALILEO [61] and modernization of the current GPS (see Chapter 6) are another motivation to search for effective synergies in a combined GNSS/UMTS-CDMA2000 architecture.

Both GPS III and GALILEO will have a direct sequence-code division multiple access (DS-CDMA) technique (with a pilot channel), as well as the air interface of the third-generation cellular communications. Because of the different frequency band allocations of GNSS and 3G communications (Figure 9.18), it is necessary to use a particular dual-frequency receiver [62, 63].

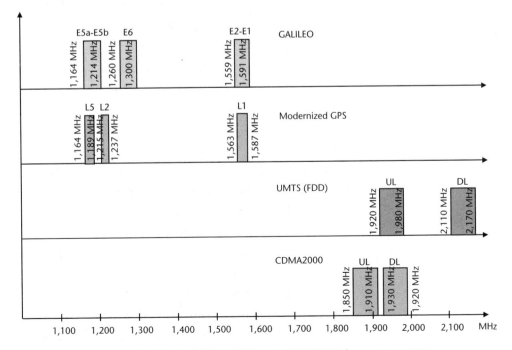

Figure 9.18 GALILEO, modernized GPS, UMTS, and CDMA2000 frequency spectra.

Even if the DS-CDMA technique is adopted both in satellite navigation and cellular mobile communications, their different objectives (i.e., global 3D positioning and timing and error-free transmission of large amounts of data, respectively) necessitates the implementation of different user equipment, mostly for code acquisition and code tracking. Generally, a stand-alone GNSS receiver has a sliding correlator for code acquisition, one for each satellite to be tracked to synchronize the received PRN code and the locally generated replica; once acquired, the satellite signal is then tracked by means of a delay-lock loop (DLL). On the other hand, mobile phones for CDMA communications use the RAKE receiver, in which a matched filter synchronizes the received PRN code and the locally generated replica. The conceptual schemes of the two receiver structures are illustrated in Figure 9.19 [60].

In [60], it is shown that a modified RAKE receiver architecture is the most suitable to meet the needs of satellite navigation and 3G communications within the same mobile user equipment.

Recently, the integration of GPS and WCDMA radio front-ends has been proposed and demonstrated [64], also highlighting the problem of intersystem (interference level from the wireless system into GPS) and intrasystem (interference levels at the GPS antenna generated internally in the GPS receiver) isolations, previously described in [65].

9.3.2 Satellite Networks

Integration of GNSS with various satellite networks is highlighted in the following sections.

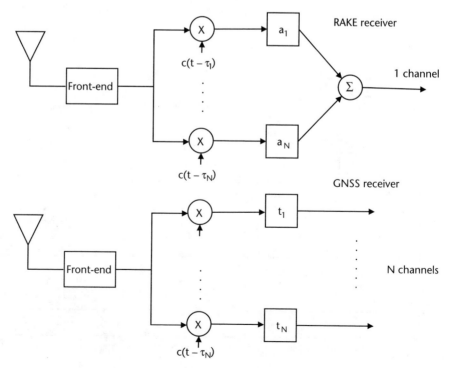

Figure 9.19 GNSS and RAKE receiver conceptual schemes.

9.3.2.1 Inmarsat

Inmarsat can be considered a pioneer of global mobile satellite communications. Formed as a maritime-focused intergovernmental organization in 1979, Inmarsat has been a limited company since 1999 and, in 2004, operated a fleet of nine (2004) geosynchronous telecommunications satellites. The coverage is almost global because the geostationary satellites cannot adequately cover the poles and high latitudes. Inmarsat has developed a number of maritime, aeronautical, and portable terminals and can provide voice, fax, Internet, data transfer, networking, videoconference, emergency services, and monitoring tracking and control services. Moreover some Inmarsat satellites carry SBAS navigation transponders both for EGNOS and WAAS. Inmarsat offers solutions for asset monitoring and tracking with two types of GPS/Inmarsat integrated terminals, called *Inmarsat mini-C* and *Inmarsat D+*. From these mobile terminals, it is possible to have packet data communication (hundreds of bytes). Messages are transmitted by a store and forward method. It takes a minimum of 5 to 10 minutes from the time of transmission until the time of reception. Some examples of these terminals are shown in Figure 9.20.

A network of hundreds of Inmarsat partners covers more than 80 countries around the world and attends to terminal sales.

Inmarsat satellites are controlled from headquarters sited in London, where the Inmarsat Group Holdings Ltd., Inmarsat's parent company, is also located, together with the International Mobile Satellite Organization (IMSO). IMSO has the task of controlling company public-service duties to support the global maritime distress and safety system (GMDSS) and satellite-aided air traffic control for the aviation community.

The Inmarsat business strategy points to carry into orbit the new Inmarsat I-4 satellites, the largest commercial communications spacecraft, which will enter service in 2005, as the backbone of the planned broadband global area network (BGAN) services. BGAN services will provide bit-rates up to 432 kbps for

Figure 9.20 GPS/Inmarsat integrated terminals.

Internet access, mobile multimedia, and other advanced applications. BGAN communications capacity will be at least 10 times superior to the current Inmarsat network [66].

9.3.2.2 Iridium

Iridium is the most challenging communication satellite system ever launched in space. The Iridium space segment is made up of 66 operational LEO satellites and 14 spares, each one capable of antisatellite links in the Ka-band and direct user communications in the L-band. The system initially foresaw 11 gateways worldwide, in order to transmit the huge predicted phone traffic. Currently the entire system uses a single commercial gateway in Tempe, Arizona. Details of the difficulties experienced by the Iridium project are outside the scope of this book. Suffice it to say that the Iridium story is very interesting and completely changed the global vision of satellite communications [67]. The coverage of Iridium is global (including the poles) and it provides voice, data, short messaging, fax, and paging from and to any location in the world. New Iridium modems with an integrated GPS feature have begun to enter the market (Figure 9.21). Iridium offers a short burst data (SBD) transmission for value-added applications, such as asset tracking and fleet management. The target services are commercial or government related, including ground or airborne applications, that require real-time transmission of position, velocity, and time.

9.3.2.3 Globalstar

Globalstar is an international multicorporate partnership for satellite telecommunication. The Globalstar system is the world's most widely used handheld satellite phone service. It is made up of 48 operational satellites in LEO orbit, plus an additional four satellites as spares. The Globalstar coverage is almost worldwide.

The satellites use an S-band (downlink) and an L-band (uplink) to communicate with the users and a C-band for feeder link. Globalstar uses a simple "bent pipe" architecture, without onboard processing, to transmit the data. The system currently offers these main services: voice, data, fax, paging, short message service, and position. The latter service is performed by time, angle, and distance information measurements from the satellite's signals, combined with an algorithm to determine geographic position in latitude and longitude. The horizontal accuracy of this service is about 10 km. Moreover, Globalstar is recently offering

Figure 9.21   GPS/Iridium integrated modem.

an interesting user terminal for asset tracking, called *AXTracker*. This integrates a low-power GPS module, a sensor's input interface, and a Globalstar simplex satellite transmitter. The simplex modem is designed to deliver packet-switched data for low-power and low-cost devices to enable communications for remote sensing and monitoring applications. The AXTracker shown in Figure 9.22 is a battery-powered rugged device for fleet management, cargo, and other mobile assets tracking.

9.3.2.4   Thuraya

Thuraya is a satellite telecommunications company founded in the United Arab Emirates (UAE) in 1997 as a private joint stock company. At the time of this writing, its shareholders are both national telecommunications societies (UAE, Qatar, Libya, Kuwait, Bahrain, Oman, Yemen, Egypt, Algeria, Sudan, Tunisia, Pakistan) and investment corporations (Abu Dhabi Investment Company (UAE), Dubai Investment PJSC (UAE), Al Murjan Trading and Industrial Company (Kingdom of Saudi Arabia), Gulf Investment Corporation (Kuwait), Deutsche Telecom Consulting GmbH (Germany), and International Capital Trading Company (UAE) [68]. It mainly offers cost-effective satellite-based mobile telephone services to nearly one-third of the globe.

The constellation is made up of two geostationary satellites positioned in geosynchronous orbit at 44°E and 28.5°E. A third satellite is being realized by Boeing Satellite Systems and is scheduled to be ready in 2005 [69]. It had been thought to expand the coverage over Asian countries. The user terminals are dual-mode and therefore can communicate with Thuraya satellites in L-band or with terrestrial GSM networks. The main services supported are voice, fax, data, short messaging, and location determination (GPS). The latter service is exploited by an integrated GPS receiver in the GSM/satellite user terminal.

The handheld terminal can indicate to the user the distance and direction of a previous stored position. This feature is very useful to navigate in large open spaces, such as deserts. The terminal itself can also perform location transmission via SMS. The vehicular terminal is mainly used for fleet management purposes. Communication is carried out using the SMSs sent via satellites or the GSM terrestrial network.

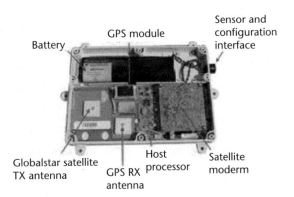

Figure 9.22   The AXTracker terminal.

9.3.2.5 Orbcomm

Orbcomm is a satellite data communication company composed of partners all around the world [Orbcomm Inc., Orbcomm Andes Caribe, Orbcomm South America, Orbcomm Europe, European Datacomm Holding N.V., Orbcomm Maghreb, SatCom International Group Ltd., Celcom (Malaysia) Berhad, Korea Orbcomm Ltd., and Orbcomm Japan Ltd.]. Orbcomm has value-added resellers (VAR) to serve customers, to which it also offers expertise and support [70].

Orbcomm provides global, packet-switched messaging capabilities. It is made up of 30 operational satellites in LEO orbit, launched into six orbital planes. The system was designed for messaging, machine to machine (M2M) communications, and position information delivery. The transmitting frequencies are in the economical VHF band. Current applications integrating GPS in the user terminal are related to tracking mobile assets such as trucks, vessels, or trains and they are working to offer infomobility services. The company Echoburst Inc. has presented a user terminal that integrated the following features:

- 16-channel GPS receiver;
- SBAS engine for WAAS, EGNOS, and MSAS;
- Gyro-based dead reckoning engine;
- GSM, GPRS, and EDGE communications capabilities;
- Orbcomm communication capabilities;
- Voice communications.

This terminal can enable advanced fleet management, both global location and tracking services, for special applications such as the security vehicle market.

9.3.2.6 QZSS

There is currently a huge demand for mass-market GPS-related products in Japan. There were about 3.8 million GPS-equipped cellular phones in Japan in 2003 [71] and about 2 million GPS-equipped car navigation units are sold annually. According to market forecasts, this figure will reach 2.7 million units sold per year. The MSAS alone will not support the required location-based services in all environments because the geostationary satellites cannot be tracked in the dense (high-rise building) areas of the Japanese metropolis. A consortium of Japanese industries, called Advanced Space Business Corporation (ASBC), including Mitsubishi Electric Corp., Hitachi Ltd., and GNSS Technologies Inc., with the assistance of the government, is currently developing a system called QZSS, for Quasi Zenith Satellite System. It is estimated that the system will be operational in 2008 and will provide a new integrated service for mobile applications based on communications, video, audio, data broadcast, positioning, and GPS augmentation. The targeted services are ITS, LBS, and obviously mobile communication in general. The innovative concept of QZSS is the choice of the orbits. The constellation will consist of three satellites moving in periodical HEOs over the Asian region. Some possible constellation configurations (Figure 9.23) are currently under study.

Satellite ground track	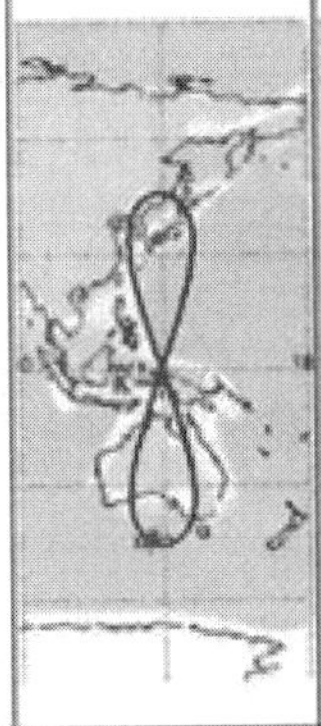	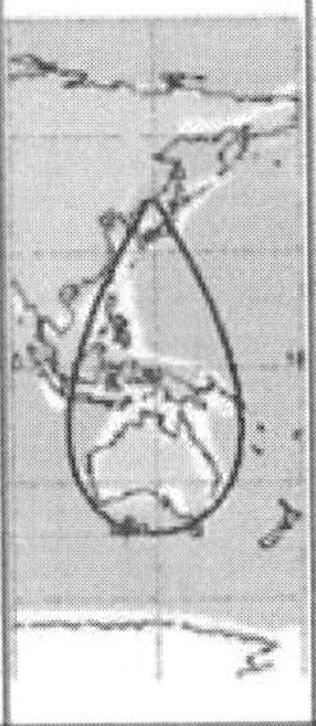	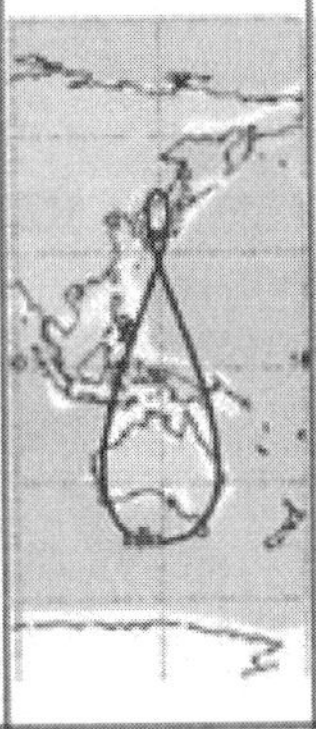		
Inclination	45°	42.5°	45°	52.6°	52.6°
Eccentricity	0	0.21	0.099	0.36	0.36
Number of satellites	3	3	3	3	4

Figure 9.23 QZSS constellations studies.

The advantage of this constellation is the high availability even for high masking angles, because of the quasi zenith position. QZSS is a clear example of integration of navigation and communications at system level for a specific targeted user community.

9.3.3 Emergency Networks

As outlined in previous chapters, satellite-based navigation systems are critical to the provision of aid and support in safety-critical and emergency applications. In terms of "integration," GNSS services can be exploited as a building block for dedicated emergency networks. A representative example for this is the *SAR service* of GALILEO, which integrates the GNSS with the emergency COSPAS-SARSAT system (see Chapter 5).

Other examples have been reported, such as a GIS/GPS/GSM-based system for *ambulance management and emergency incident handling* in the Greek prefecture of Attica [72]. The integrated system is expected to operate in the National Center of Immediate Assistance (ETAK in Greek), which deals with emergency medical incidents by coordinating and routing ambulances to appropriate hospitals and offering medical care to patients during their transport to hospitals. A further system, the Taiwan *E-Vanguard for Emergency* (EVE), provides help in rescue and first-aid work in rescue operations [73]. EVE consists of rescuers equipped with wearable PDAs, medics with laptops plus GPS, a rescue command center controlled by leaders of rescue and medical teams, and a monitor subsystem that includes physical sensors, such as an oximeter and electron-cardiograph, provided with wireless transmission for the injured. The various parts of the system communicate with each other by means of Bluetooth. In Figure 9.24, the system architecture is displayed. A GPS module—EverMore GM-X205, which supports the NXMEA-0183 output format defined by

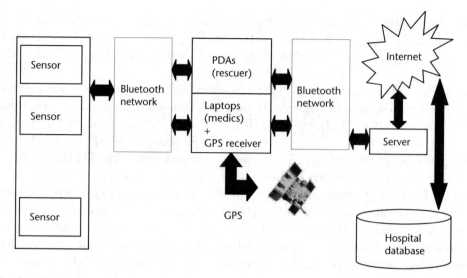

Figure 9.24   Architecture of the EVE system.

NMEA—is installed in every laptop used by outdoor medics. It calculates the medics' positions, using the information from satellites, and transfers them to the server, where they can be displayed on its electronic map, using Bluetooth. Further applications of the system are envisaged in the future, such as homecare for patients with chronic illnesses.

Another example of integration is provided by the Japanese air-to-air surveillance system for helicopter safety and efficient operations, which consists of a GPS-based position sensor, VHF data link (VDL), and cockpit display for traffic information (DTI) [74]. The use in the air-to-air case had been demonstrated successful, achieving large improvement—a "see and avoid" capability—on the situational awareness of surrounding traffic. Further applications have then been conceived, such as air-to-ground data link communications and a specific airspace surveillance ground monitoring system, in particular for *emergency medical service* (EMS) *helicopters* (Figure 9.25).

A further integrated application is reported in [75], concerning the development of a GPS position location transceiver unit in a *GPS-based land search environment* (GLASE) *system*, which is aimed at providing exact position information of search and rescue personnel in the field to the Waverley Ground Search and Rescue (WGSR) tactical search management team in Nova Scotia.

An emergency-related application currently under development is the *SCORE (EGNOS-based Coordinated Operational Emergency and Rescue) project*, which will set up emergency call positioning (E-112, the European version of the U.S. E-911) and rescue force guidance services during accidents or natural disasters [76].

In the military field, the *combat search and rescue* (CSAR) *program* is worth mentioning. The core of CSAR is the combat survivor/evader locator (CSEL) radio, a communications system that provides a survivor/evader—such as a downed pilot—with some integrated tools. These tools are precision GPS-based geoposition and navigation data, the way over-the-horizon (OTH) secure data

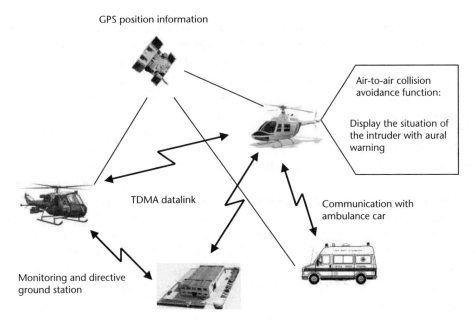

GPS position information

Air-to-air collision avoidance function:

Display the situation of the intruder with aural warning

TDMA datalink

Communication with ambulance car

Monitoring and directive ground station

Figure 9.25   Architecture of the EMS helicopter integrated system.

communications to Joint Search and Rescue Centers (JSRC), OTH beacon operation and line-of-sight voice communications, and swept tone beacon capabilities [77, 78]. Another application concerns the *use of robots* in hostile-to-human situations, such as across a landscape contaminated by radiation, biological warfare, or chemical spills, in an earthquake area rippled with aftershocks, and through a minefield or crossfire zone. Backpack-sized robots, carried to the scene by individual soldiers or emergency personnel, cannot transport all mission payloads that might be necessary to investigate an area of interest. In this context, the *leader/follower robot concept* has been introduced: one robot traces out a safe path via user control and autonomous navigation, then other robots can be sent along the same path using the leader/follower behavior. The lead robot can also be "told" to follow its own path backwards to perform a retro-traverse. The use of GPS is naturally envisaged for both the leading and following robots to plot their way into and out of danger [79].

References

[1]   Schiller, J., *Mobile Communications*, Reading, MA: Addison-Wesley, 2000.

[2]   Prasad, R., and M. Ruggieri, *Technology Trends in Wireless Communications*, Norwood, MA: Artech House, 2003.

[3]   Kayton, M., and W. R. Fried, *Avionics Navigation Systems*, 2nd ed., New York: John Wiley & Sons, 1997.

[4]   Galati, G., "Detection and Navigation Systems/Sistemi di Rilevamento e Navigazione," Rome, Italy: Texmat, 2002.

[5]   Parkinson, B. W., and J. J. Spilker Jr., (eds.), *Global Positioning System: Theory and Applications*, Vol. 163 and 164, Washington, D.C.: American Institute of Aeronautics and Astronautics, 1996.

[6]   Roth, L., et al., "Status of U.S. Loran Evaluation and Integrated GPS/Loran Development Program—Ability of a Modern Loran System to Augment and Backup GNSS," *GNSS 2003, The European Navigation Conference*, Graz, Austria, April 2003.

[7]   van Willigen, D., G. W. A. Offermans, and A. W. S. Helwig, "EUROFIX: Definition and Current Status," *IEEE Position Location and Navigation Symposium*, April 1998, pp. 101–108.

[8]   van Willigen, D., et al., "Improving GPS UTC Service by LORAN-C/Eurofix, *GNSS 2003, The European Navigation Conference*, Graz, Austria, April 2003.

[9]   Tetley, L., and D. Calcutt, *Electronic Navigation Systems*, 3rd ed., Boston, MA: Butterworth-Heinemann, 2001.

[10] http://www.rednova.com/news/stories/1/2004/10/14/story122.html.

[11] Kleusberg, A., "Comparing GPS and GLONASS," *GPS World*, Vol. 1, No. 6, November/December 1990, pp. 52–54.

[12] Daly, P., et al., "Frequency and Time Stability of GPS and GLONASS Clocks," *International Journal of Satellite Communications*, Vol. 9, No. 1, 1991, pp. 11–22.

[13] Anodina, T. G., "Status and Prospects for the Development of the GLONASS Satellite Navigation System," *Proc. FAANS(II)/4-WP/47, ICAO*, Montreal, Canada, September 1993.

[14] Langley, R. B., "GLONASS Review and Update," *GPS World*, Vol. 8, No. 7, July 1997, pp. 46–51.

[15] Misra, P., et al., "Receiver Autonomous Integrity Monitoring (RAIM) of GPS and GLONASS," *Navigation*, Vol. 40, Spring 1993, pp. 87–104.

[16] Bazlof, Y. A., et al., "GLONASS to GPS: A New Coordinate Transformation," *GPS World*, Vol. 10, No. 1, January 1999, pp. 54–58.

[17] *"GALILEI—Navigation Systems Interoperability Analysis," Gali-THAV-DD080*, October 2002.

[18] "GALILEO and GPS Will Navigate Side by Side: EU and US Sign Final Agreement," *IP/04/805*, Brussels, Belgium, June 2004.

[19] "Progress in GALILEO-GPS Negotiations," *IP/04/173*, Brussels, Belgium, 2004.

[20] Doherty, J., "Directions 2004 Part 1," *GPS World*, December 1, 2003.

[21] Gibbons, G., "Interoperability: Not So Simple," *GPS World*, December 1, 2003.

[22] Issler, J. L., et al., "GALILEO Frequency & Signal Design," *GPS World*, June 1, 2003.

[23] Godet, L., et al., "GALILEO Spectrum and Interoperability Issue," *Proc. GNSS2003*, Graz, Austria, April, 2003.

[24] Belli, R. G., "GPS and GALILEO—Capabilities and Compatibility," *Proc. European Satellites for Security Conference*, Brussels, Belgium, June 2002.

[25] Turner, D. A., "Compatibility and Interoperability of GPS and GALILEO: A Continuum of Time, Geodesy, and Signal Structure Options for Civil GNSS Services," in M. Rycroft, (ed.), *Satellite Navigation Systems: Policy, Commercial and Technical Interactions*, Boston, MA: Kluwer, 2003, pp. 85–102.

[26] Karner, J., "Status of GPS-GALILEO Cooperation—A US Perspective," *Munich Satellite Navigation Summit*, March 2004, http://www.munich-satellite-navigation-summit.org/Summit2004.

[27] Hilbrecht, H., "GALILEO Institutional and International Issues," *Munich Satellite Navigation Summit*, March 2004, http://www.munich-satellite-navigation-summit.org/Summit2004.

[28] Vultaggio, M., A. Greco, and L. Russo, "The NETNAV Integrated Navigation System/Il Sistema Integrato di Navigatzione NETNAV," *Atti IIN*, No. 172, September 2003, pp. 46–55.

[29] El-Rabbany, A., *Introduction to GPS: The Global Positioning System*, Norwood, MA: Artech House, 2002.

[30] Kaplan, E. D., (ed.), *Understanding GPS: Principles and Applications*, Norwood, MA: Artech House, 1996.

[31]   Curey, R. K., et al., "Proposed IEEE Inertial Systems Terminology Standard and Other Inertial Sensor Standards," *Proc. of Position Location and Navigation Symposium,* Monterey, CA, April 2004.

[32]   Barbour, N., "Inertial Components—Past, Present and Future," *Proc. AIAA GNC Conf.,* Montreal, Canada, August 2001.

[33]   Barbour, N., and G. Schmidt, "Inertial Sensor Technology Trends," *IEEE Sensors Journal,* Vol. I, No. 4, December 2001.

[34]   Lawrence, A., *Modern Inertial Technology: Navigation, Guidance, and Control,* 2nd ed., New York: Springer-Verlag, 1998.

[35]   Dalla Mora, M., A. Germani, and C. Manes, "Introduzione Alla Teoria Dell'identificazione Dei Sistemi," *Edizioni EUROMA-La Goliardica,* Rome, Italy, 1997.

[36]   Hajiyev, C., and M. A. Tutucu, "Development of GPS Aided INS Via Federated Kalman Filter," *Proc. of Int. Conf. on Recent Advances in Space Technologies,* Istanbul, Turkey, 2003.

[37]   Hide, C., T. Moore, and M. Smith, "Adaptive Kalman Filtering Algorithms for Integrating GPS and Low Cost INS," *Proc. of Position Location and Navigation Symposium,* Monterey, CA, April 2004.

[38]   Song, C., B. Ha, and S. Lee, "Micromachined Inertial Sensors," *Proc. of the 1999 IEEE/RSJ Int. Conf. on Intelligent Robots and Systems,* Kyongju, Korea, October 1999.

[39]   Yazdi, N., F. Ayazi, and K. Najafi, "Micromachined Inertial Sensors," *Proc. of the IEEE,* Vol. 86, No. 8, August 1998.

[40]   "Revision of the Commission's Rules to Ensure Compatibility with Enhanced 911 Emergency Calling Systems," *Report and Order and Further Notice of Proposed Rulemaking,* FCC, Washington, D.C., June 1996.

[41]   *FCC Acts to Promote Competition and Public Safety in Enhanced Wireless 911 Services,* WT Rep. 99-27, FCC, Washington, D.C., September 15, 1999.

[42]   U.S. Federal Communication Commission Web Site About E-911, http://www.fcc.gov/911/enhanced.

[43]   National Emergency Number Association (NENA) Web Site About E-911, http://www.nena.org/Wireless911/index.htm.

[44]   Zhao, Y., "Mobile Phone Location Determination and Its Impact on Intelligent Transportation Systems," *IEEE Trans. on Intelligent Transportation Systems,* Vol. 1, No. 1, March 2000, pp. 55–64.

[45]   Zhao, Y., "Standardization of Mobile Phone Positioning for 3G Systems," *IEEE Communications Magazine,* Vol. 40, No. 7, July 2002.

[46]   Christie, J., et al., "Development and Deployment of GPS Wireless Devices for E911 and Location Based Services," *IEEE Position, Location and Navigation Symposium,* Palm Springs, CA, April 15–18, 2002.

[47]   Lopes, L., E. Viller, and B. Ludden, "GSM Standards Activity on Location," *Proc. of IEEE Colloquium on Novel Methods of Location and Tracking of Cellular Mobiles and Their System Application* (Ref. No.1999/046), London, England, May 17, 1999.

[48]   Lee, D. J. Y., and W. C. Y. Lee, "Optimize CDMA System Capacity with Location," *Proc. of 12th IEEE Int. Symp. on Personal, Indoor and Mobile Radio Communications 2001,* San Diego, Calif., September 30–October 3, 2001, pp. D144–D148.

[49]   Lee, D.-S., and Y.-H. Hsueh, "Bandwidth-Reservation Scheme Based on Road Information for Next-Generation Cellular Networks," *IEEE Trans. on Vehicular Technology,* Vol. 53, No. 1, January 2004, pp. 243–252.

[50]   Chiu, M. H., and M. A. Bassiouni, "Predictive Schemes for Handoff Prioritization in Cellular Networks Based on Mobile Positioning," *IEEE Journal on Selected Areas in Communications,* Vol. 18, Issue 3, March 2000, pp. 510–522.

[51] Feng, S., and C. L. Law, "Assisted GPS and Its Impact on Navigation in Intelligent Transportation Systems," *Proc. of IEEE 5th International Conference on Intelligent Transportation Systems*, Singapore, September 3–6, 2002.

[52] Djuknic, G. M., and R. E. Richton, "Geolocation and Assisted GPS," *IEEE Computer Magazine*, Vol. 34, No. 2, February 2001, pp. 123–125.

[53] Enge, P., R. Fan, and A. Tiwari, "GPS Reference Networks' New Role—Providing Continuity and Coverage," *GPS World*, July 2001, pp. 38–45.

[54] "Location Technologies for GSM, GPRS and UMTS Networks," SnapTrack—A QUALCOMM Company, Location Technologies White Paper X2, 2003; http://www.cdmatech.com/resources/pdf/location_tech_wp_1-03.pdf.

[55] Martin-Escalona, I., F. Barcelo, and J. Paradells, "Delivery of Nonstandardized Assistance Data in E-OTD/GNSS Hybrid Location Systems," *Proc. of the 13th IEEE Int. Symp. on Personal, Indoor and Mobile Radio Communications*, Lisbon, Portugal, September 2002, pp. 2347–2351.

[56] Nissani, D. N., and I. Shperling, "Cellular CDMA (IS-95) Location, A-FLT (Assisted Forward Link Trilateration) Proof-of-Concept Interim Results," *Proc. of the 21st IEEE Convention of the Electrical and Electronic Engineers in Israel*, Tel-Aviv, Israel, April 11–12, 2000, pp. 179–182.

[57] http://www.cdmatech.com/solutions/products/gpsone_cdma.jsp.

[58] EMILY Project Web site, http://www.emilypgm.com.

[59] "Cellular/GNSS Hybrid Module Specification and Interfaces," EMILY.IST-2000-26040, Deliverable 9, September 9, 2003.

[60] Heinrichs, G., R. Bischoff, and T. Hesse, "Receiver Architecture Synergies Between Future GPS/GALILEO and UMTS/IMT-2000," *Proc. of 56th IEEE Vehicular Technology Conference*, Vancouver, Canada, September 24–28, 2002.

[61] Issler, J.-L., et al., "Galileo Frequency & Signal Design," *Galileo's World*, June 1, 2003.

[62] Hekmat, T., and G. Heinrichs, "Dual-Frequency Receiver Technology for Mass Market Applications," *Proc. of the Institute of Navigation ION GPS-01 International GPS Conference*, Salt Lake City, UT, September 2001.

[63] Eissfeller, B., et al., "Real-Time Kinematic in the Light of GPS Modernization and Galileo," *Galileo's World*, October 1, 2002.

[64] Spiegel, S. J., and I. I. G. Kovacs, "An Efficient Integration of GPS and WCDMA Radio Front-Ends," *IEEE Trans. on Microwave Theory and Techniques*, Vol. 52, No. 4, April 2004, pp. 1125–1131.

[65] Spiegel, S., et al., "Improving the Isolation of GPS Receivers for Integration with Wireless Communication Systems," *Proc. IEEE Radio Frequency Integrated Circuits Symposium*, Philadelphia, PA, June 2003.

[66] Inmarsat Web site, http://www.inmarsat.com.

[67] Chen, C., "Iridium: From Punch Line to Profit?" *Fortune*, Vol. 146, No. 4, 2002, p 42.

[68] Thuraya Web Site, http://www.thuraya.com.

[69] From the Official Web Site for the Ministry of Information and Culture in the United Arab Emirates, http://www.uaeinteract.com/news/default.asp?ID=257.

[70] Orbcomm Web Site, http://www.orbcomm.com.

[71] Petrovsky, I. G., "QZSS—Japan's New Integrated Communication and Positioning Service for Mobile Users," *GPS World*, June 2003.

[72] Derekenaris, G., et al., "An Information System for Effective Management of Ambulances," *Proc. of 13th IEEE Symposium on Computer-Based Medical Systems*, Houston, TX, 2000.

[73] Chen, S. C., et al., "E-Vanguard for Emergency—A Wireless System for Rescue and Healthcare," *Proc. of IEEE Enterprise Networking and Computing in Healthcare Industry*, Santa Monica, CA, June 2003, pp. 29–35.

[74]   Yokota, M., Y. Kubo, and N. Kuraya, "Some VDL Applications for Helicopter Safety and Efficient Operation," *Proc. of the 21st IEEE Digital Avionics Systems Conference*, Irvine, CA, October 2002, pp. 10.A.3-1-11.

[75]   Bower, D., et al., "Design and Development of the GLASE Position Location Transceiver," *Proc. of 1994 Canadian Conference on Electrical and Computer Engineering*, Halifax, Canada, September 1994.

[76]   "ALCATEL to the Rescue with Project SCORE," *GPS World*, May 1, 2004.

[77]   Luccio, M., "Guiding Weapons, Finding Soldiers," *GPS World*, July 31, 2002.

[78]   "Interstate to Support CSEL Rescue Radios," GPS Inside-December 2002, *GPS World*, December 1, 2002.

[79]   Hogg, R., "Send in the 'Bots,'" *GPS World*, August 1, 2002.

CHAPTER 10
Open Issues and Perspectives

10.1 A Step Ahead for Aviation: The Free Flight

In the last decade air traffic has been characterized by a significant yearly growth that resulted in a dramatic increase in delays and compromised safety and efficiency in the air traffic environment. This growth is predicted to continue at a rate of 2.5% per year over the next few years [1, 2]. Therefore, radical measures are required to deal with the negative impacts of the growth in air traffic. This has led to a necessity to design and develop a new ATM concept—called *Free Flight*—to optimize the use of the airspace.It is a concept for future ATM that aims at increasing system capacity, improving safety standards, and fully using capacity resources to provide a significant enhancement in the efficiency of ATM.

The Free Flight concept has been around since the early 1990s, but the FAA officially showed its interest when it established a special task force to study and develop the project, named the *Radio Technical Commission for Aeronautics* (RTCA) *Free Flight Task Force*. The Free Flight concept was defined in 1995 in the official report of the RTCA as [3]:

> A safe and efficient flight operating capability under Instrument Flight Rules (IFR) in which the operators have the freedom to select their path and speed in real time.
> —RTCA, 1995

NASA is participating in the research on free flight by developing a far-term concept of operations, the *distributed air/ground traffic management* (DAG TM). In particular, the DAG TM Concept Element 5 (CE 5) [4] specifies operations in the en route and terminal-transition flight domains. In addition DAG TM CE 5 presents a new category of flight operations, the *autonomous flight rules* (AFR). AFR-equipped aircraft are authorized to dynamically plan and execute their preferred trajectories without coordinating with the ground-based air traffic service (ATS) provider. Therefore, AFR-equipped aircraft flight crews are responsible for traffic separation and conformance to operational constraints established by ATS providers to preserve special-use airspace and handle traffic flow in crowded terminal areas. The ATS provider does not get involved in AFR procedures in en route and terminal-transition domains under normal operations, whereas he or she still provides IFR services to nonautonomous aircraft.

An important advantage envisioned in AFR adoption is the accommodation of a traffic volume higher than that manageable by a ground-based IFR system.

It is important to highlight that Free Flight is not "free for all" since air traffic restrictions are still imposed on the aviation crew/pilots. However, the restrictions are limited in extent and duration to deal with any identified problem [3]. Intervention from a ground-based ATS provider is limited to four situations in order to:

- Preclude exceeding airport capacity;
- Ensure separation;
- Prevent unauthorized flight through special-use airspace;
- Ensure safety of flight.

In particular, under nominal conditions, DAG TM CE 5 states that the interactions between autonomous aircraft and the ATS provider include the *traffic flow management* (TFM), which is performed by time-based arrival metering: the ATS provider delivers required-time-of-arrival (RTA) clearances and crossing restrictions at metering fixes to autonomous aircraft flight crew, which has to accept them. In the remaining time until the fix crossing by the aircraft, there are minimum dealings between the ATS provider and the aircraft [5].

The Free Flight concept gives pilots more freedom and independence from the ground-based ATM system, while maintaining the traditional protection afforded under IFR by using advanced technology [3].

Therefore, Free Flight is intended to either minimize or remove air traffic restrictions that slow down the ground-based ATM system, allowing pilots free routing to find the most optimal trajectories to fly.

Free Flight in its mature state is expected to significantly improve ATM, resulting in cost-effective air traffic service provision.

The FAA Free Flight program was organized in three phases (Table 10.1 [6]).

FFP1 started in October 1998 and ended in December 2002. New ATM tools were developed and tested for operational feasibility under Free Flight. The FFP1 results were largely positive, and this encouraged the FAA to continue the Free Flight program. The follow-on phase, FFP2, is currently underway, with the objective of improving and expanding the FFP1 capabilities. In addition, a new data link system, called the Controller Pilot Data Link Communications (CPDLC), will be developed to enhance air-ground communications. The third phase, FFP3, is expected to start in 2006 and will complete the rollout of software, hardware, and the implementation of other results from the first two phases [6]. The success of FFP1 also had the positive effect that Europe decided to participate actively and contribute to the Free Flight project. The Italian air traffic service, ENAV, in cooperation with several European avionic agencies and companies, initiated a large five-year program, called the Mediterranean Free Flight (MFF), in 1999. MFF aims to study the feasibility of Free Flight in the Mediterranean area. The program started in 2000 and consists of two operative phases, running from 2000–2003 and 2004–2005, respectively [7].

Table 10.1 Free Flight Program Schedule

	Start	End
Free Flight Phase 1 (FFP1)	1998	2002
Free Flight Phase 2 (FFP2)	2003	2005
Free Flight Phase 3 (FFP3)	2006	2015

In this context, GNSSs and their advanced capabilities will play a key role in achieving the Free Flight goals. In fact, since the Free Flight concept provides pilots with free routing capability, the need of autonomous and accurate positioning is a paramount and critical requirement to achieving the highest level of safety and efficiency in air traffic management.

It is worth mentioning that humans will remain the most critical element in the use of new technologies, procedures, and equipment. In a future Free Flight environment, aviation crew/pilots should be able to electronically see and manage any critical situation. In particular, the intercommunication capability between aircraft represents a key element in conflict detection and resolution. In this respect, a GNSS-based technology is currently under development, named Air-to-Air Automatic Dependent Surveillance-Broadcast (ADS-B). ADS-B will allow equipped aircraft to automatically broadcast their position to any "listeners" (other aircraft) in the local vicinity [8].

10.2   HAP-Based Integrated Networks

As already mentioned in Chapter 8, HAPs are unmanned long-endurance aircraft, operating at altitudes comprised between 15 and 30 km. They are less expensive, more adaptable, and closer to the ground than artificial satellites. In addition, their flight path can be controlled from the ground, hence be guidable to serve different regions [9].

On the other hand, as outlined in Chapter 5 and 8, the *GALILEO local component* will play a key role in the future navigation system. The GALILEO local component will enhance GALILEO navigation performance and capabilities to fulfill particular markets or user requirements that the GALILEO global component alone could not accomplish. The GALILEO local component consists of GALILEO local elements and is a part of the overall GALILEO definition. The GALILEO Programme forecasts the development of a few and selected experimental local elements.

HAPs can play several roles as a GNSS local component (Figure 10.1).

The following logical separation between the functionality that can be fulfilled by HAPs helps classify this role:

- *Layer 1*: enhancement of the RNP (i.e., accuracy, availability, integrity, and continuity);
- *Layer 2*: enhancement of the communication capabilities to provide advanced navigation and localization services.

While this separation is mainly logical, sometimes it also does not correspond to a physical separation of components in the HAP payload.

With regard to the first layer, a single HAP can provide enhancement of the RNP on a local area (ranging from hundreds of meters to hundreds of kilometers) because of its limited visibility range, whereas a constellation of HAPs is needed to operate on continental areas. The following list summarizes possible usages of HAPs as a system to improve the RNP functionality [9]:

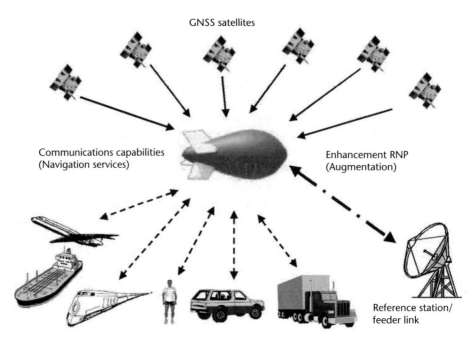

Figure 10.1   HAPs as a GNSS local component.

- *Stratospheric Ranging and Monitoring Station.* HAPs can be included in a terrestrial RIMS network acting as moving RIMSs.
- *Integrity messages broadcasting* (increase of integrity). HAPs can broadcast integrity messages computed by a terrestrial RIMS or autonomously computed (acting as RAIM).
- *Ranging correction broadcasting* (increase of accuracy). HAPs can include corrections for satellite ephemeris, clock correction, and ionospheric propagation errors.
- *Additional ranging signal* (increase of availability and continuity).
- *Local assisted navigation service* (increase of availability and continuity, especially in difficult environments). This concerns the techniques to reduce the TTFF and receiver tracking threshold. A real-time two-way communication is needed to deliver pseudorange and ephemeris information to users.

About layer two, the integration between navigation and mobile communication networks enables the provision of GNSS services with added value. The integration with the following communication modes can be conceived:

- Broadcast and multicast communications;
- Multipoint-to-point communications;
- Two-way narrowband communications;
- Two-way broadband communications.

In this framework, HAPs can act as a local component for LBS. HAPs could also be used for connectionless multipoint-to-point communications. Only considering navigation-related services, the following new cost-effective way to exploit fleet management and traffic control could be proposed: mobile users (i.e., trucks or private cars) can transmit their position (computed by GALILEO), speed, and identification code to the HAP by using a low-cost transmitter-only terminal; data are forwarded to the control center for traffic control or fleet management. The same system infrastructure could be used for antitheft or emergency messages (e.g., automatic incident detection). The advantages of such a system with respect to a satellite constellation (e.g., ORBCOMM) are threefold:

- The cost of the service (for each HAP) could be shared with other communication services.
- A single HAP could be a complete and functional system locally, for instance by exploiting fleet management in a metropolis.
- The improved link budget reduces the level of transmit power as well as cost and size of user terminal.

The advantages of HAP with respect to a terrestrial communication system for fleet management (e.g., SMS) are due to the connectionless communications, which allows a low-cost and high-rate message transmission.

The integration of HAPs and mobile networks can bring analogous advantages in the large set of navigation services that use two-way broadband and narrowband communications, ranging from advanced driver assistance systems to updating digital maps to personal navigation. Navigation-related services should make use of the same networks used for Internet and voice services, thus sharing the costs.

The Galilei studies [10] (a cluster of studies under the Growth thematic programme of the 5th Research Framework Programme of the EU) defined a number of integration scenarios with 3G and 4G terrestrial communication networks. From this, commercial applications of the "local component" concept are foreseen. The integration scenarios propose a framework for guaranteeing service agreements for advanced telematics and infomobility applications. Finally, this framework could drive the price of selling policies regarding "public utilities services," "best-effort commercial services," and "guaranteed commercial services."

As outlined in Chapter 8, HAPs can be considered a key building block of future communications-navigation integrated scenarios.

10.3   B3G Mobile Networks

The current research and development (R&D) scenarios for *beyond third generation* (B3G) mobile networks envision a technological step up in order to provide new applications and new services. Nevertheless, any attempt to define new communication means or new multimedia services clashes with the needs for the mobility of the user. Therefore, future mass-market user terminals must be light, handheld, low power, and possibly low cost to be accepted by the people. These requirements are challenging and, at the same time, appealing for user terminal manufacturers. In fact, future user terminals will be wearable, have projectable or flexible displays, projectable keyboards, and voice recognition capabilities, and be light powered and multistandard. Most of the innovation for B3G will be in the terminals. The mobile network role in the B3G scenario foresees seamless integration between different networks with different standards, protocols, and technologies. All these features will enable higher bandwidth for users and global seamless coverage: the motto is high-speed wireless Internet, voice, and mobile videoconferencing every time and everywhere. The B3G scenarios foresee the improvement of previous services and applications that can be exploited with low performances from 3G networks. Moreover, 3G is slowly spreading worldwide and its new services seem not only driven by consumer requirements but rather by technology availability. However, 3G was conceived as a step to understand more deeply user needs and requirements.

Recalling the history of mobile communication, we remember that the 1G network (about 20 years ago) provided only a voice service with low quality of service and low coverage. The terminals were mostly vehicular at the beginning and then "rugged" portable ones with low autonomy. What followed was a revolution in mobile communication.

Nowadays, satellite navigation terminals can mainly provide route guidance service; mass-market user terminals are mostly vehicular and portable ones have low autonomy and low coverage (e.g., cannot work indoors or in urban canyons). A new revolution is at hand, and the drivers will be twofold:

- Integration with B3G networks for advanced LBS;
- Global mobile localization.

Perhaps the most common question that mobile users ask each other at the beginning of a call is: "Where are you?" This means that localization is a need and this need is currently not fully met by existing technologies. The user-centric approach is very common in the B3G scenario [11] and it envisions that the network could know our needs and satisfy them even before we ask. This concept is often implemented by the network through the study of our actions in order to define a user profile. This model cannot be effective and efficient if the network cannot know where we are. In the near future, B3G networks will propose a discount on our new favorite suit when approaching a store, will find a vegetarian restaurant nearby, will say where our best friend is and how to reach him or her and, finally, the table at the restaurant and the food will be ready exactly at the moment we arrive. All these services need the knowledge of user positions and the integration of this information into B3G networks. Moreover, the services have to be available whenever and wherever required. Satellite navigation is only a component of the global mobile localization concept. A current hot R&D topic is the study of indoor positioning using the technologies developed for communication purposes, including wireless LANs [12–14], Bluetooth [15–18], A-GPS [19], and various combinations [20]. Satellite navigation systems will be used for outdoor (or semi-indoor) localization only and integration with communication networks will enable seamless global mobile localization. Every technology will give the raw range or angular measurement, and the network will calculate and provide to the user the positioning information. There will be a full intersystem interoperability because the B3G networks will be designed to have centralized control of positioning information.

10.4   Digital Divide

Digital divide is commonly meant as the gap existing among different communities about their capability to access new technologies.

> People lack many things: jobs, shelter, food, health care, and drinking water. Today, being cut off from basic telecommunications services is a hardship almost as acute as these other deprivations, and may indeed reduce the chances of finding remedies to them.
>
> —Kofi Annan, U.N. Secretary General, 1999

Each new technology introduced in the developed world should take into account environmental impact, ethics, and the digital divide. A major concern with respect to access to the new technologies is that the distance between the developed and developing countries is increasing day by day. The reasons for this are manifold, including:

- The near-total lack of an internal market for new technologies into the third world;
- New technologies are expensive and often need costly infrastructure development;
- Social, environmental, cultural, and educational differences often make the natives unable to use the same technological equipment used in the first world.

The role that space-based technologies can play in bridging the digital divide could be of primary importance. Space infrastructures could offer services to wide areas without any major added cost. There are many examples of that: telecommunications geostationary satellite coverage often includes developing areas; Earth observation LEO satellite coverage always includes developing countries; and satellite navigation systems are intrinsically global positioning systems.

Nevertheless positioning in rural and wide areas, such as deserts or mountains, is indeed a useful service. The issue of digital divide is now the focus of attention from governments, politicians, and other institutions. Technology makers should be encouraged to play the important role of designing "ad hoc" services to cater to the needs of the developing world.

10.5   B2G GNSS

Readers familiar with the mobile communications world might remember what transpired in the period 1999–2000. While GSM (i.e., a 2G mobile system) was gathering global success and all engineering and technological efforts were being directed at developing 3G mobile networks, some experts in the field were starting to meet, brainstorming and trying to figure out what the 4G mobile communication systems and devices could look like or, even, could mean [21, 22].

The navigation world is experiencing a similar situation: GPS (i.e., 1G-GNSS) has experienced global success. Major activities are ongoing to develop 2G-GNSS (i.e., GALILEO, GPS III). Naturally, this will be followed by initial activities centered around a 3G-GNSS concept. Improvement in quantity and quality of services as well as in user terminal features, compactness, and cost are easy predictions for the third generation.

Surprises and novelties may come from the space segment. Two main directions can be identified in the progress of satellite constellations: pioneering of novel frequency ranges, where high capacity is available for transferring large volumes of data at high rate; and development of highly connected satellite constellations using interorbit links (IOL) to improve system efficiency. Interesting experimentation of unexplored portions of the extremely high frequency (EHF) range is being proposed. In particular, propagation and communications experiments are being developed by the Italian Space Agency (ASI) in the W-band (75–110 GHz) using the 85- GHz range for the ground-to-satellite link and the 75-GHz range for the opposite direction [23, 24]. The W-band also seems promising for IOL, due to the almost free-space behavior of the channel that avoids the atmosphere-related supplementary attenuation.

The vision of a 3G-GNSS passes, for instance, through an augmented 2G-GNSS (i.e., a 2.5 G-GNSS), where the augmentation links are developed at a high rate in the W-band. The 3G-GNSS constellation could be equipped with IOL capabilities, perhaps with a W-band IOL.

In a more general vision, the navigation satellite constellation may be seen as a building block of a complex, integrated architecture composed of different layers: terrestrial, HAP, and satellite (LEO, MEO, and GEO) layers [25]. The multilayer architecture is depicted in Figure 10.2. Connections may be both intralayer and interlayer.

In this context, a GNSS augmentation can be seen as an interlayer link between the MEO layer (GNSS) and, for instance, the GEO layer (augmentation satellites). Many possible interlayer augmentations of the GNSS can be conceived. In general,

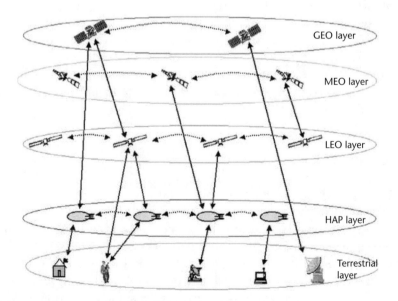

Figure 10.2 The multilayer system architecture.

integration cases can be foreseen where systems operate in different layers and/or extend over more than one layer.

In all cases in the future vision, some specific elements will play fundamental roles: interconnectivity, rate/frequency range, and integration. Any projection we wish to create will include a cocktail of these elements and the differences among the possible projections will be given by the different percentage of the ingredients in the mixture.

10.6 Navigation Services Penetration

Chapter 8 provided a wide illustration of GNSS applications, involving very different areas and entering into sectors unthinkable a few years ago. This section presents some figures of GNSS penetration in various markets. During the last few years the GNSS industry has seen continuous growth in the number of firms that use satellite navigation to create value-added services, ranging from GIS and land surveys to LBS and ITS. This trend toward the provision of consumer services was forecast in many studies carried out in the 1990s [26–28], later confirmed by recent research (mainly before selective availability was deactivated). It was reported in [29] that in terms of revenues, the aviation market for GPS products had a growth of 10%, the marine market 11%, the military and timing markets both just under 25%, and the land market just over 24%. Approximately 62% of the total North American GPS market revenue was from land applications. With the turning off of SA the improved accuracy has encouraged the implementation of new navigation services. A further shift toward service provision is due to the modernization of GPS and the European navigation activities, with the introduction of EGNOS in 2004 and with the GALILEO constellation scheduled to become operational in 2008. In 2000 the European Commission launched the Galilei Project to survey preliminary analysis and definition of requirements for the development of GALILEO. One of the main results was that the earlier view that GALILEO and GPS would be competitors had to be reverted, indicating that the default receiver will be a combined GPS and GALILEO one. Forecasts envisage that more than 2.5 billion GALILEO-enabled receivers will be in use by 2020, with the use of combined GPS/GALILEO receivers being the norm by 2012 [30] (Figure 10.3).

The Galilei Project also estimated that the product market will begin to saturate after 2015, leaving its dominant position, whereas the service market will become the driving force up to 2020. In the latter, mobile phone and road transport navigation services [vehicle telematics, road charging, and advanced driver assistance systems (ADAS)] will dominate the others. GALILEO, operating as a second constellation, will improve availability and overall performance, hence increasing user perception of service levels and, consequently, bolstering market growth. It is estimated that the number of GNSS-enabled mobile phones will reach two billion by 2020 (Figure 10.4) [31]. This is equivalent to a penetration of 70% of the overall mobile phone market (which will consist of about 2.9 billion handsets) [31].

As far as road transport is concerned, it is envisioned that 495 million vehicles will have a GNSS terminal by 2020 [30]. This includes mass-market vehicles (i.e.,

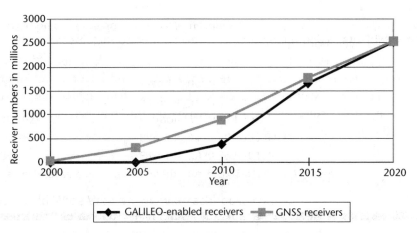

Figure 10.3   GNSS receivers in the period 2000–2020.

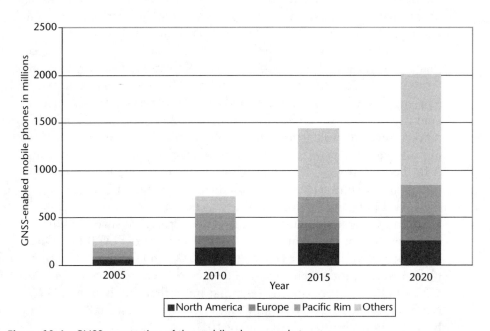

Figure 10.4   GNSS penetration of the mobile phone market.

cars) and commercial vehicles (i.e., trucks, buses, and light commercial vehicles), as shown in Figures 10.5 and 10.6 [31].

In the car market, North America and Europe will have the leading role, with a similar number of cars equipped with GNSS receivers by 2020.

However, in Europe, there are a larger number of cars and a stronger need for information services due to higher level of congestion in traffic. These features make the rate of penetration of GNSS receivers in the European car market higher than in North America, as shown in Figure 10.5 by the greater slope of line relative to Europe. On the contrary, in the commercial vehicle market, North America leads (note that sports utility vehicles, or SUVs, are included in the number of commercial vehicles). It is assessed that the percentage of penetration of these systems by 2020 in

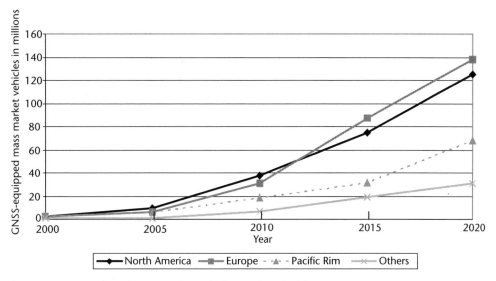

Figure 10.5   Number of mass-market vehicles equipped with GNSS.

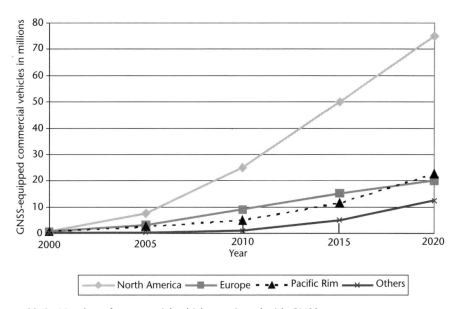

Figure 10.6   Number of commercial vehicles equipped with GNSS.

the whole commercial vehicles market will be 50% (absolute value 130 million). With regard to the use of GALILEO signals onboard commercial vehicles, it has to be noted that, in this market segment the replacement cycle of trucks or buses that have already adopted a GPS receiver is longer than that in the mobile phone segment. It is predicted that 16 million and 114 million commercial vehicles will be fully equipped with combined GPS/GALILEO receivers by 2010 and 2020, respectively [31].

Looking at the whole market size as number of sold GNSS receivers, the percentage of LBS, including the mobile phone sector and the road transport sector, will rise to about 92% (absolute values 2 billion and 495 million, respectively) by 2020, leaving 8% (absolute value 180 million) to the other sectors [31].

From the gross revenue point of view, comparing the two "bull" market segments, the GNSS receiver market revenue for road vehicles is not as inferior to that of mobile handsets as one would expect considering only the number of sold units. This fact is due to the higher costs of a GNSS terminal for vehicles because it includes various peripherals, such as CD players, large color screens, and so on. Therefore, the figures are established at 76 billion euros for the former and at 94 billion euros for the latter (once again by 2020) [30].

10.7 Conclusions

The matter described throughout the nine previous chapters should have convinced even the most skeptic reader that the number of current and future applications of satellite navigation systems is small compared with their great and—perhaps—largely unexpressed potential impact on society.

In this chapter, we have attempted to highlight current applications and imagine (or simply whisper) some of the possible application scenarios of GNSS based on the underpinning scientific/technical background and achievements to date. The reader should find the content of this book useful as a foundation to the wide and interesting world (current and future) of space-based navigation and communications.

References

[1]   *Annual Report of the Council 2003*, ICAO, 2004.

[2]   *Annual Report Eurocontrol 2003*, Eurocontrol, 2004.

[3]   *Final Report of the RTCA Board of Directors' Select Committee on Free Flight*, Radio Technical Commission for Aeronautics, Technical Report, Washington, D.C., January 18, 1995.

[4]   "DAG-TM Concept Element 5 En Route Free Maneuvering Operational Concept Description," NASA Advanced Air Transportation Technologies Project Office, 2002.

[5]   Krishnamurthy, K., et al., "Autonomous Aircraft Operations Using RTCA Guidelines for Airborne Conflict Management," *22nd Digital Avionics Systems Conference*, Indianapolis, IN, October 12–16, 2003.

[6]   Post, J., and D. Knorr, "Free Flight Program Update," *5th USA/Europe Air Traffic Management R&D Seminar*, Budapest, Hungary, June 2003.

[7]   Iodice, L., G. Ferrara, and T. Di Lallo, "An Outline About the Mediterranean Free Flight Programme," *3rd USA/Europe Air Traffic Management R&D Seminar*, Naples, Italy, June 2000.

[8]   "National Airspace System Capital Investment Plan Fiscal Years 2002–2006," Federal Aviation Administration, April 2001.

[9]   Avagnina, D., et al., "Wireless Network Based on High-Altitude Platforms for the Provision of Integrated Navigation/Communication Services," *IEEE Communications Magazine*, Vol. 40, No. 2, February 2002.

[10] "Summary of Local Element Customization Study," *Galilei*, July 2003.

[11] Prasad, R., and M. Ruggieri, *Technology Trends in Wireless Communications*, Norwood, MA: Artech House, 2003.

[12]   Kotanen, A., et al., "Positioning with IEEE 802.11b Wireless LAN," *14th IEEE Proc. on Personal, Indoor and Mobile Radio Communications*, Vol. 3, September 2003, Beijing, China, pp. 2218–2222.

[13]   Kitasuka, T., T. Nakanishi, and A. Fukuda, "Wireless LAN-Based Indoor Positioning System WiPS and Its Simulation," *IEEE Pacific Rim Conference on Communications, Computers and Signal Processing*, Vol. 1, Victoria, British Columbia, Canada, August 2003, pp. 272–275.

[14]   Vossiek, M., et al., "Wireless Local Positioning," *IEEE Microwave Magazine*, Vol. 4, No. 4, December 2003, pp. 77–86.

[15]   Kotanen, A., et al., "Experiments on Local Positioning with Bluetooth," *Proc., International Conference on Information Technology: Coding and Computing (Computers and Communications)*, Las Vegas, NV, April 2003, pp. 297–303.

[16]   Hallberg, J., M. Nilsson, and K. Synnes, "Positioning with Bluetooth," *10th International Conference on Telecommunications*, Vol. 2, Tahiti, February–March 2003, pp. 954–958.

[17]   Thongthammachart, S., and H. Olesen, "Bluetooth Enables Indoor Mobile Location Services," *57th IEEE Semiannual Vehicular Technology Conference*, Vol. 3, Jeju, Korea, April 2003, pp. 2023–2027.

[18]   Anastasi, G., et al., "Experimenting an Indoor Bluetooth-Based Positioning Service," *Proc., 23rd International Conference on Distributed Computing Systems Workshops*, Providence, RI, May 2003, pp. 480–483.

[19]   Kedong, W., et al., "GpsOne: A New Solution to Vehicle Navigation," *Position Location and Navigation Symposium*, Monterey, CA, April 2004, pp. 341–346.

[20]   Dhruv, P., J. Ravi, and E. Lupu, "Indoor Location Estimation Using Multiple Wireless Technologies," *14th IEEE Proc. on Personal, Indoor and Mobile Radio Communications*, Vol. 3, Beijing, China, September 2003, pp. 2208–2212.

[21]   Prasad, R., and M. Ruggieri, (eds.), Special Issue on "Future Strategy for the New Millenium Wireless World," *Wireless Personal Communications*, Vol. 17, Nos. 2–3, June 2001, pp. 149–153.

[22]   Prasad, R., and M. Ruggieri, (eds.), Special Issue on "Designing Solutions for Unpredictable Future," *Wireless Personal Communications*, Vol. 22, No. 2, August 2002, pp. 103–108.

[23]   Ruggieri, M., et al., "The W-Band Data Collection Experiment of the DAVID Mission," *IEEE Trans. on Aerospace and Electronic Systems*, Special Section on "The DAVID Mission of the Italian Space Agency," Vol. 38, No. 4, October 2002, pp. 1377–1387.

[24]   De Fina, S., M. Ruggieri, and A. V. Bosisio, "Exploitation of the W-Band for High-Capacity Satellite Communications," *IEEE Trans. on Aerospace and Electronic Systems*, Vol. 39, No. 1, January 2003, pp. 82–93.

[25]   Cianca, E., and M. Ruggieri, "SHINES: A Research Program for the Efficient Integration of Satellites and HAPs in Future Mobile/Multimedia Systems," *Proc. WPMC*, Yokosuka, Japan, October 2003.

[26]   Kaplan, E. D. (ed.), *Understanding GPS: Principles and Applications*, Norwood, MA: Artech House, 1996.

[27]   "Differential GPS Markets in the 1990s, a Cross-Industry Study," KV Research, 1992.

[28]   "Global Positioning System—Market Projections and Trends in the Newest Global Information Utility," The International Trade Administration, Office of Telecommunications, U.S. Department of Commerce, September 1998.

[29]   "GPS Report," Frost & Sullivan, May 2000.

[30]   "The Galilei Project—GALILEO Design Consolidation," European Commission, Esys Plc., Guildford, United Kingdom, August 2003.

[31]   Styles, J., N. Costa, and B. Jenkins, "In the Driver's Seat—Location-Based Services Power GPS/GALILEO Market Growth," *GPS World*, October 1, 2003.

List of Acronyms

A-FLT advanced forward link trilateration

A-GPS assisted-GPS

A/D analog-to-digital

AAIM aircraft autonomous integrity monitoring

ABAS aircraft-based augmentation systems

ADF/NDB automatic direction finder/nondirectional beacon

ADAS advanced driver assistance systems

ADS-B Air-to-Air Automatic Dependent Surveillance-Broadcast

AHRS Attitude and Heading Reference System

AHS Automated Highway System

AII Accuracy Improvement Initiative

AKF adaptive Kalman filter

AMCS alternate master control station

AOA angle of arrival

APV approach for vertical guidance

ARNS Aeronautical Radio Navigation Services

ARRC Alaska Railroad Corporation

ARTEMIS Advanced Relay Technology Mission

ASBC Advanced Space Business Corporation

AS antispoofing

ASQF Application Specific Qualification Facility

ATM air traffic management

AUV autonomous underwater vehicles

B3G beyond third generation

BGAN broadband global area network

BIH Bureau International de l'Heure

BIPM Bureau International des Poids et Mesures

BPSK binary phase shift keying

BS base station

BSC base station controller

BTS Bureau of Transportation Statistics

BTS84 BIH Terrestrial System 1984

C-WAAS Canadian WAAS

CARAT Computer And Radio Aided Train control system

CENC China-Europe global Navigation satellite system technical training and cooperation Center

CAS collision avoidance system

CC composite clock

CCDS Consultative Committee for Definition of Second

CCF central control facility

CDGPS conventional DGPS

CDMA code division multiple access

CDU control display unit

CGSIC Civil GPS Service Interface Committee

COSPAS-SARSAT Cosmicheskaya Sistyema Poiska Avariynich Sudov–Search and Rescue Satellite-Aided Tracking

CPDLC controller pilot data link communications

CPF central processing facility

CRV crew return vehicle

CS commercial service

CSAR combat search and rescue

CSEL combat survivor/evader locator

CTP conventional terrestrial pole

CTRS conventional terrestrial reference systems

DECT Digital European Cordless Telecommunications

DEM Digital Elevation Model

DGNSS differential GNSS

DGPS differential GPS

DGRS differential GPS reference station

DL design life

DLL delay-lock loop

DMA Defense Mapping Agency

DME distance measuring equipment

DNSS Defense Navigation Satellite System

DOC Department of Commerce

DoD Department of Defense

DOP dilution of precision

DORIS Doppler Orbitography Radiopositioning Integrated by Satellite

DORIS Differential Ortho-Rectification Imaging System

DOT Department of Transportation

DSP digital signal processor

DTG dynamically tuned gyroscope

DTI   display for traffic information

E-OTD   enhanced observed time difference

EATCHIP   European Air Traffic Control Harmonization and Integration Program

EC   European Commission

ECAC   European Civil Aviation Conference

ECEF   Earth-centered Earth-fixed

ECI   Earth-centered inertial

EDGE   enhanced data rates for GSM evolution

EGNOS   European Geostationary Navigation Overlay Service

ELT   emergency location terminals

EMS   emergency medical service

ENT   EGNOS network time

EPIRB   Emergency Position Indicating Radio Beacons

EPN   EUREF Permanent Network

EPS   electronic payment services

ERNP   European Radio Navigation Plan

ERTMS   European Rail Traffic Management System

ESA   European Space Agency

ETCS   European Train Control System

ETML   European traffic management layer

ETRS89   European TRS 1989

EU   European Union

EUROCAE   European Organization for Civil Aviation Electronics

EVE   E-Vanguard for Emergency

EWAN   EGNOS wide area communication network

FAA   Federal Aviation Administration

FANS   Future Air Navigation Systems

FCC   Federal Communications Commission

FDE   fault detection and exclusion

FDI   Fault Detection and Isolation

FEC   forward error correction

FFP1   Free Flight Phase 1

FFP2   Free Flight Phase 2

FFP3   Free Flight Phase 3

FGC   Ferrocarriles Generalitat Catalunya

FHWA   Federal Highway Administration

FOC   full operational capability

FOG   fiber-optic gyroscopes

FRA   Federal Railroad Administration

FRS   Federal Radionavigation Systems

FTA   Federal Transit Administration

GA   ground antenna

GA   general aviation

GAGAN   GPS and GEO augmented navigation

GAN   global area network

GBAS   ground-based augmentation system

GCC   GALILEO Control Center

GDOP   geometric dilution of precision

GEO   geostationary Earth orbit

GERAN   GSM EDGE Radio Access Network

GIAC   GPS Interagency Advisory Council

GIS   geographic information system

GLAS   GALILEO locally assisted service

GLASE   GPS-based Land Search Environment system

GLONASS   GLObal'naya NAvigatsionnaya Sputnikovaya Sistema

GMDSS   Global Maritime Distress and Safety System

GNSS   Global Navigation Satellite System

GOC   GALILEO Operating Company

GPRS   general packet radio service

GPS   global positioning system

GRACE   gravity recovery and climate experiment

GRI   group repetition interval

GRS   deodetic reference system

GSM   global system for mobile communications

GSOS   GALILEO satellite-only services

GST   GALILEO system time

GSS   GALILEO sensor station

GSSB   GALILEO System Security Board

GTRF   GALILEO terrestrial reference frame

GUS   GALILEO uplink station

HAL   horizontal alert limit

HAP   high-altitude platform

HDOP   horizontal dilution of precision

HARM   high-speed antiradiation missile

HEO   highly elliptical orbit

HHA   harbor/harbor approach

HPL   horizontal protection limit

HRG   hemispherical resonant gyroscope

I/O   input/output

IAG   International Association of Geodesy

IARP   Italian Antarctic Research Program

IAU   International Astronomical Union

ICAO   International Civil Aviation Organization

ICD   interface control document

ICS   integrated control system

IEEE/AES   Institute of Electrical and Electronics Engineers/Aerospace and Electronics Systems Society

IERS   International Earth Rotation and Reference Systems Service

IFOG   interferometric fiber-optic gyroscope

IFR   instrument flight rules

IGIS   Indoor Guidance and Information System

IGS   International GPS Service for Geodynamics

ILS   Instrument Landing System

IMO   International Maritime Organization

IMSO   International Mobile Satellite Organization

IMU   inertial measurement unit

INS   Inertial Navigation System

IOC   initial operational capability

IOL   interorbit link

IRM   IERS reference meridian

IRP   IERS reference pole

IRU   inertial reference unit

ISA   inertial sensor assembly

ISL   intersatellite link

ISS   International Space Station

ITRF   international terrestrial reference frame

ITRS   International Terrestrial Reference System

ITRS-PC   TRS Product Center

ITS   Intelligent Transportation System

ITS-JPO   Intelligent Transportation Systems Joint Program Office

IUGG   International Union of Geodesy and Geophysics

ITU   International Telecommunication Union

JPO   Joint Program Office

JSRC   Joint Search and Rescue Center

JU   joint undertaking

KOMPSAT-1/2   Korea Multipurpose Satellite

LAAS   Local Area Augmentation System

LADGPS   local area DGPS

LBS   location-based service

LEC   local exchange carrier

LEO   low Earth orbit

LF   low frequency

LFSR   linear feedback shift register

LGF   LAAS ground facility

LIDAR   LIght Detection And Ranging

LLR   lunar laser ranging

LNAV/VNAV   lateral navigation/vertical navigation

LNM   local notice to mariners

LOP   line of position

LORAN-C   LOng RAnge Navigation

LOS   line-of-sight

LPV   lateral precision vertical guidance

LRS   long range terrestrial navigation system

LS   landing system

LUT   local user terminal

M2M   machine to machine

MARAD   Maritime Administration

MC   master clock

MCC   mission control center

MCS   master control station

MEMS   Microelectromechanical system

MEO   medium Earth orbit

MF   medium frequency

MFF   Mediterranean Free Flight

MLS   microwave landing system

MMD   mean mission duration

MMS   mobile mapping system

MMAE   multiple model adaptive estimation

MOEMS   Micro-optical-electromechanical system

MOPS   minimum operational performance standards

MS   monitor station

MSAS   MTSat Satellite Augmentation System

MSC   mobile switching center

MSK   minimum shift keying

MSN   monitor station network

MTSat   Multifunction Transport Satellites

MUS   mission uplink station

MVDS   microwave vehicle detection system

NAS   National Airspace System

NASA   National Aeronautics and Space Administration

NANU   notice advisory to navigation users

NAVCEN   NAVigation CENter

NAVSTAR   NAVigation Satellite Timing And Ranging

NDGPS   networked DGPS

NGS   National Geodetic Survey

NHTSA   National Highway Traffic Safety Administration

NIMA   National Imagery and Mapping Agency

NIS   Navigation Information Service

NLES   Navigation Land Earth Station

NMEA   National Marine Electronics Association

NNSS   Navy Navigation Satellite System

NOAA   National Oceanic and Atmospheric Administration

NOTAM   Notice to Airmen

NPA   nonprecision approach

NSWCDD   Naval Surface Warfare Center Dahlgren Division

NTS   navigation technology satellite

OCS   operational control segment

OS   open service

OTH   over-the-horizon

PACF   Performance Assessment and system Checkout Facility

PDA   personal digital assistant

PDOP   position dilution of precision

PHM   passive hydrogen maser

PL   pseudolite

PNU   precision navigation upgrade

PPP   public-private partnership

PPS   precise positioning service

PPT   personalized public transit

PRIMA   Italian Multi-Applications Reconfigurable Platform

PRN   pseudo random noise

PRS   public regulated service

PSAP   public safety answering point

PVT   position, velocity, and timing

QPSK   quadrature phase shift keying

QZSS   Quasi Zenith Satellite System

RAIM   receiver autonomous integrity monitoring

Rb   Rubidium

RC   ranging code

RCC   Rescue Coordination Center

RCS   Regional Control Station

RDC   R&D center

RF   radio frequency

RIMS   Ranging and Integrity Monitoring Station

RINEX   Receiver IndepeNdent EXchange

RLG   ring laser gyroscope

RLSP   return link service provider

RNP   required navigation performance

RNSS   Radio Navigation Satellite Service

RRA   reference receiver antenna

RS   reference station

RSPA   Research and Special Programs Administration

RTCA   Radio Technical Commission for Aeronautics

RTCM SC-104   Radio Technical Commission for Maritime Services, Special Committee 104

RTK   real-time kinematic

SA   selective availability

SAR   search and rescue

SARPS   Standard And Recommended PracticeS

SBAS   satellite-based augmentation system

SBD   short burst data

SEAD   suppression of enemy air defenses

SFD   satellite failure detection

SIS   signal in space

SLBM   submarine launched ballistic missile

SLR   satellite laser ranging

SLSDC   St. Lawrence Seaway Development Corporation

SLSS   Secondary Lines Signaling System

SMS   short message service

SNAS   Satellite Navigation Augmentation System

SNUGL   Seoul National University GPS Lab

SoL   safety of life

SOLAS   safety of life at sea

SPS   standard positioning service

SRS/MRS   short/middle range terrestrial radio navigation system

TACAN   Tactical Air Navigation

TAI   international atomic time

TANS   Trimble Advanced Navigation Sensor

TCAR   three-carrier phase ambiguity resolution

TCXO   Temperature-Compensated Crystal Oscillators

TDOA   time difference of arrival

TDOP   time dilution of precision

TDRSS   Tracking and Data Relay Satellite System

TEC   total electron content

TEN   Trans European transport Network

TIMATION   TIMe/navigATION

TMC   traffic message channel

TOA   time of arrival

TRF   terrestrial reference frame

TRS   terrestrial reference system

TSO   technical standard order

TT&C   tracking, telemetry and command

TTA   time to alert

TTFF   time-to-first-fix

UE   user equipment

UEE   user equipment error

UERE   user equivalent range error

UHF   ultra high frequency

UMTS   Universal Mobile Telecommunications System

URE   user range error

USAF   U.S. Air Force

USCG   United States Coast Guard

USNO   United States Naval Observatory

UTRAN   Universal Terrestrial Radio Access Network

UT   universal time

UT   user terminal

UT1   universal time 1

UTC   universal time coordinated

VAL   vertical alert limit

VAR   value-added reseller

VBA   vibrating beam accelerometer

VDB   VHF data broadcast

VDL   VHF data link

VDOP   vertical dilution of precision

VeRT   vehicular remote tolling

VFR   visual flight rules

VHF   very high frequency

VLBI   very long baseline interferometry

VLF   very low frequency

VMS   variable message signs

VOR   very high frequency omni range

VPEMS   vehicle performance and emissions monitoring system

VRS   virtual reference station

VTS   vessel traffic systems

WAAS   wide area augmentation system

WAD   wide area differential

WADGPS   wide area DGPS

WCDMA   wideband code division multiple access

WGS-84   World Geodetic System – 1984

W-LAN   wireless local area network

WMS   wide area master station

WRS   wide area reference station

About the Authors

Marina Ruggieri graduated in 1984 from the University of Roma with a degree in electronics engineering. She has worked at FACE-ITT and GTC-ITT in Roanoke, Virginia, in the High Frequency Division (1985–1986); as a research and teaching assistant at the University of Roma Tor Vergata (RTV) (1986–1991); and as an associate professor in telecommunications at the University of L'Aquila (1991–1994). Since November 2000, she has worked as a full professor in telecommunications at the RTV (Department of Electronics Engineering), teaching digital signal processing, information, and coding.

Ms. Ruggieri is the director of an M.Sc. program on advanced satellite communications and navigations systems at RTV. She teaches satellite systems and applications at the International Teledoc program for doctorate students on the CNIT terrestrial-satellite integrated network. Her research mainly concerns space communications and navigation systems (in particular, satellites), as well as mobile and multimedia networks.

Ms. Ruggieri is a member of the Technical and Scientific Committee of the Italian Space Agency (2004–2007). In addition, she is the principal investigator of various research programs, including: satellite scientific communications missions (*DAVID, WAVE*) of ASI; a national research program (*CABIS*) on CDMA integrated mobile systems (2000–2002) and on satellite-HAP integrated networks for multimedia applications (*SHINES*), cofinanced by MIUR (2002–2004). She coordinates the RTV unit in various European projects, including: EU FP6 IP *MAGNET* (My personal Adaptive Global NET); EU ASIA LINK *EAGER-NetWIC* (Euro-Asian Network for Strengthening Graduate Education and Research in Wireless Communications); EU Network *NEXWAY*; GALILEO JU 1st Call—July 2003: *VERT* (VEhicular Remote Tolling); and in the ASI program on V-band payloads (*TRANSPONDERS*). In 1999, she was appointed member of the Board of Governors of the IEEE AES Society (2000–2002) and reelected for 2003–2005.

Ms. Ruggieri is editor of the *IEEE Transactions on Aerospace and Electronic Systems* (AES) for "Space Systems" and chair of the IEEE AES Space Systems Panel. Since 2002, she has served as cochair of Track 2 "Space Missions, Systems, and Architecture" of the AES Conference. She was reappointed to the IEEE Judith A. Resnik Award Committee for 2004 and has been a member of TPC for PLANS 2004. She is also a member of the editorial board of *Wireless Personal Communications—An International Journal* (Kluwer). She was awarded the 1990 Piero Fanti International Prize and nominated for the Harry M. Mimmo Award in 1996 and the Cristoforo Colombo Award in 2002.

In addition, Ms. Ruggieri has authored about 180 papers, international journals/transactions and proceedings of international conferences, book chapters, and books. She is also an IEEE Senior Member (S'84-M'85-SM'4).

Ramjee Prasad received his bachelor of science degree in engineering from the Bihar Institute of Technology in Sindri, India, and his master's of science degree in engineering and Ph.D. from Birla Institute of Technology (BIT) in Ranchi, India, in 1968, 1970, and 1979, respectively.

He joined BIT as a senior research fellow in 1970 and became an associate professor in 1980. While he was with BIT, Dr. Prasad supervised a number of research projects in the area of microwave and plasma engineering. From 1983–1988, he was with the University of Dar es Salaam (UDSM) in Tanzania, where he became a professor of telecommunications in the Department of Electrical Engineering in 1986. At UDSM, he was responsible for the collaborative project Satellite Communications for Rural Zones with Eindhoven University of Technology in the Netherlands. From February 1988 to May 1999, he was with the Telecommunications and Traffic Control Systems Group at DUT, where he was actively involved in wireless personal and multimedia communications (WPMC). He was the founding head and program director of the Center for Wireless and Personal Communications (CWPC) of International Research Center for Telecommunications—Transmission and Radar (IRCTR).

Since June 1999, Dr. Prasad has been with Aalborg University, where currently he is the director of the Center for Teleinfrastruktur (CTIF) and holds the chair of wireless information and multimedia communications. He is the coordinator of European Commission Sixth Framework Integrated Project MAGNET (My personal Adaptive Global NET). Dr. Prasad was involved in the European ACTS project FRAMES (Future Radio Wideband Multiple Access Systems) as a DUT project leader. He is also a project leader of several international, industrially funded projects.

Dr. Prasad has published more than 500 technical papers, contributed to several books, and has authored, coauthored, and edited 13 books, all published by Artech House: *CDMA for Wireless Personal Communications*; *Universal Wireless Personal Communications*; *Wideband CDMA for Third Generation Mobile Communications*; *OFDM for Wireless Multimedia Communications*; *Third Generation Mobile Communication Systems*; *WCDMA: Towards IP Mobility and Mobile Internet*; *Towards a Global 3G System: Advanced Mobile Communications in Europe, Volumes 1 and 2*; *IP/ATM Mobile Satellite Networks*; *Simulation and Software Radio for Mobile Communications*; *Wireless IP and Building the Mobile Internet*; *Multicarrier Techniques for 4G Mobile Communications*; *OFDM for Wireless Communications Systems*; and *Technology Trends in Wireless Communications*. His current research interests lie in wireless networks, packet communications, multiple-access protocols, advanced radio techniques, and multimedia communications.

Dr. Prasad has served as a member of the advisory and program committees of several IEEE international conferences. He has also presented keynote speeches and delivered papers and tutorials on WPMC at various universities, technical institutions, and IEEE conferences. He was also a member of the European cooperation in the scientific and technical research (COST-231) project dealing with the evolution of land mobile radio (including personal) communications as an expert for the Netherlands, and he was a member of the COST-259 project.

Dr. Prasad was the founder and chairman of the IEEE Vehicular Technology/Communications Society Joint Chapter, Benelux Section, and is now the honorary chairman. In addition, he is the founder of the IEEE Symposium on Communications and Vehicular Technology (SCVT) in the Benelux, and he was the symposium chairman of SCVT'93. Dr. Prasad is currently the Chairman of IEEE Vehicular Technology Communications/Information Theory Society Joint Chapter, Denmark Section.

In addition, Dr. Prasad is the coordinating editor and editor-in-chief of the *Kluwer International Journal on Wireless Personal Communications* and a member of the editorial board of other international journals, including the *IEEE Communications Magazine* and *IEE Electronics Communication Engineering Journal*. He was the technical program chairman of the PIMRC'94 International Symposium held in The Hague, the Netherlands, from September 19–23, 1994 and also of the Third Communication Theory Mini-Conference in Conjunction with GLOBECOM'94, held in San Francisco, California, from November 27–30, 1994. He was the conference chairman of the 50th IEEE Vehicular Technology Conference and the steering committee chairman of the second International Symposium WPMC, both held in Amsterdam, the Netherlands, from September 19–23, 1999. In addition, he was the general chairman of WPMC'01, held in Aalborg, Denmark, from September 9–12, 2001. He is the general chairman of the International Wireless Summit (IWS 2005), to be held in Aalborg, Denmark, on September 17–22, 2005.

Dr. Prasad is also the founding chairman of the European Center of Excellence in Telecommunications, known as HERMES. He is a fellow of IEE, a fellow of IETE, a senior member of IEEE, a member of the Netherlands Electronics and Radio Society (NERG), and a member of IDA (Engineering Society in Denmark). Dr. Prasad is also an advisor to several multinational companies.

Index

Recent Titles in the Artech House Mobile Communications Series

John Walker, Series Editor

For further information on these and other Artech House titles,
including previously considered out-of-print books now available through our In-Print-Forever®
(IPF®) program, contact:

Artech House
685 Canton Street
Norwood, MA 02062
Phone: 781-769-9750
Fax: 781-769-6334
e-mail: artech@artechhouse.com

Artech House
46 Gillingham Street
London SW1V 1AH UK
Phone: +44 (0)20 7596-8750
Fax: +44 (0)20 7630-0166
e-mail: artech-uk@artechhouse.com

Find us on the World Wide Web at: www.artechhouse.com